Programming Mobile Robots with Aria and Player

Amanda Whitbrook

Programming Mobile Robots with Aria and Player

A Guide to C++ Object-Oriented Control

 Springer

Dr. Amanda Whitbrook
University of Nottingham
School of Computer Science
IMA Research Group
Wollaton Road
Nottingham NG8 1BB
UK
amw@cs.nott.ac.uk

ISBN 978-1-84882-863-6   ISBN 978-1-84882-864-3 (eBook)
DOI 10.1007/978-1-84882-864-3
Springer London Dordrecht Heidelberg New York

British Library Cataloguing in Publication Data
A catalogue record for this book is available from the British Library

Library of Congress Control Number: 2009940604

Cover design: eStudioCalamar, Figueres/Berlin

Printed on acid-free paper

Springer is part of Springer Science+Business Media (www.springer.com)

This book is dedicated to my family; my son Charlie, my mother Josephine, my sister Squirrel, and the memories of my late brother Steven and late father Ray.

Preface

This book is intended as a comprehensive guide to object-oriented C++ programming and control of the Pioneer class of robots made by MobileRobots Inc. It covers both the native API (ARIA, supplied by the manufacturer for use with all their classes of robot), and the popular and more generic open-source Player server, which can be used with many different makes and models. Hence, although the book is written around the Pioneer robots in particular, the techniques and principles demonstrated are applicable to a wide range of other mobile robots currently in use in academic and industrial robot labs around the world.

The aim is to provide a text that can be used for the practical teaching of object-oriented programming with real robots, and also support researchers using Player and ARIA in their labs. The reader will learn how to install the necessary software, troubleshoot common problems, set up the files needed to describe their robot configuration, and will rapidly be able to get started with the task of creating their own control programs.

The text assumes some prior knowledge of object-oriented concepts since the main focus is instructing the user in the use of the ARIA API and the Player C++ client library. However, the instructions here are given primarily by example and in such a way that the object-oriented concepts themselves are also implicitly explained. Readers completely new to object-oriented programming should therefore have no problems with understanding the text and should find themselves easily getting to grips with object-oriented principles as well as learning how to program their robots.

The book is divided into six chapters. Chapter 1 provides some background information about Pioneer robots and their control including the various client-server programming architectures that can be adopted, the robot devices present and the software that is available to support them. It also quickly covers installation of the ARIA API and various other MobileRobots resources such as ACTS software (ActivMedia Color Tracking System), MobileSim (the ARIA simulator) and Mapper3Basic (software for creating navigation maps). In addition, it explains how to install Player and its simulator Stage. Chapter 2 presents detailed information on the use of the ARIA API for robot programming, showing how to connect to and control the robot and each of its devices. Chapter 3 is concerned with use of the MobileRobots resources installed in Chapter 1, i.e. ACTS, MobileSim and Mapper3Basic, and Chapter 4 rounds off the ARIA section of the book by explaining how to create and use subclasses with ARIA. Programming with the Player C++ client library is the subject of Chapter 5, and as with Chapter 2, comprehensive details about connecting to and controlling the robot and each of its devices are supplied. Chapter 6 describes the use of Player's Stage simulator and explains how to create world files and configuration files to define virtual robots, their device set-ups and their environments.

The ARIA and Player sections of the book are both fully supported by sample programs, but the reader is also directed to the online supporting materials at http://extras.springer.com, where more detailed and complex programs are available. These additional programs are intended to integrate all of the techniques presented, and they are explained further in the Appendix section.

Finally, please note that this guide is concerned with installing and using ARIA and Player software on Linux-based operating systems only, since Player is not compatible with Windows operating systems.

Nottingham, May 2009   *Amanda Whitbrook*

Acknowledgements

Thanks to my Robot Lab colleagues - Phil Birkin, Robert Oates, Jamie Twycross, Jon Garibaldi and Uwe Aickelin at the University of Nottingham, School of Computer Science.

Thanks also to all those who regularly post replies to questions on the Player / Stage mailing list, in particular, the Player / Stage developers who include Brian Gerkey, Andrew Howard, Nate Koenig, Richard Vaughan, Fred Labrosse, Geoffrey Biggs, Toby Collet and Radu Bogdan Rusu. Special thanks to Geoffrey Biggs at the University of Auckland for helpful advice and information concerning use of the Pioneer 5D arm with the Player server.

Thanks to MobileRobots Inc. for sorting out queries about the Pioneers.

Contents

Chapter 1
Introduction and Installations

1.1 The Client-Server Paradigm

Pioneer P3-DX robots act as the server in a client-server environment. The low level details of mobile robotics are managed by servers embodied in the operating system software (ARCOS, AROS, P2OS) of the robot's micro-controller [14]. The client software that provides the high level control must run on a computer connected to the micro-controller. This can either be the on-board PC that communicates with it directly through a serial connection, or via a remote networked PC, which requires a server program to be running on the robot PC, providing the communication link between the remote PC and the micro-controller. The first scenario is shown diagrammatically in Figure 1.1 below.

1.2 Software for Pioneer Robot Control

There are currently two software packages that can be used to write high level control programs for the Pioneer 3 robots. These are:

- **The ARIA API**
  This software is provided by the Pioneer manufacturers, MobileRobots Inc. (formerly ActivMedia) and can be used to control any of their models, e.g. AmigoBots, PeopleBots and Pioneers. ARIA stands for ActivMedia Robotics Interface for Application and it is an object-oriented Applications Programming Interface (API), written in C++ and intended for the creation of intelligent high-level client-side software. It is essentially a library for C++ programmers. In addition, a programmer's own "action" classes may inherit from the base ArAction class. These classes run in their own thread with the robot's current action being determined by an action resolver, see Chapter 4. This facilitates easy creation of a subsumption-like architecture, although this methodology does not have to be followed. ARIA is therefore technically architecture independent.

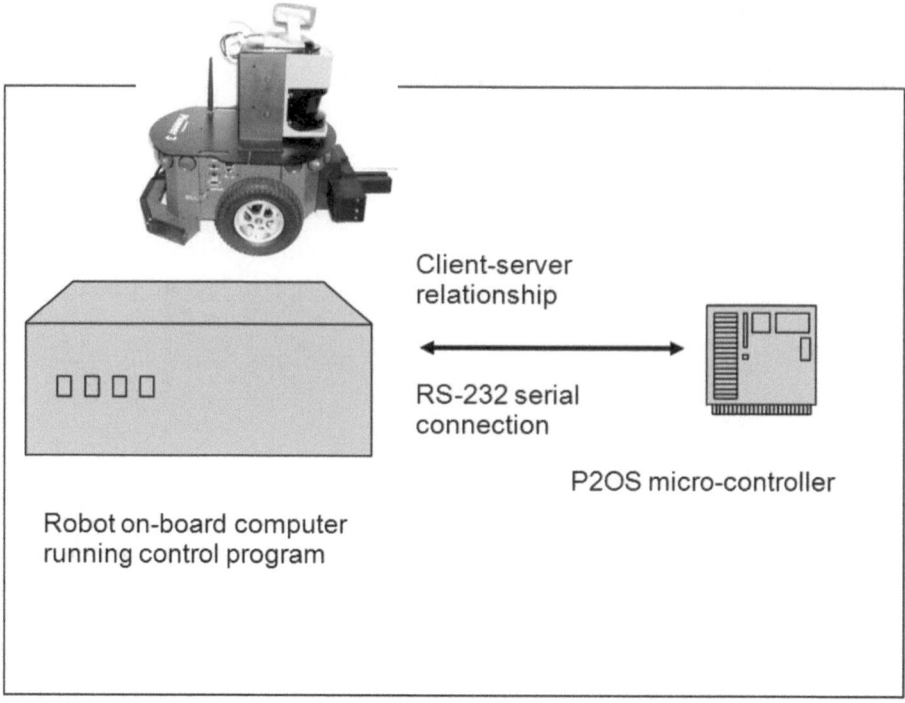

Fig. 1.1 Controlling the robot from the on-board PC

ARIA client programs can run from the robot on-board PC, communicating with the micro-controller through an RS-232 serial link, see Figure 1.1. In addition, client programs can run on a remote PC. In this scenario (see Figure 1.2) the client program requests services from a server program that runs on the robot PC. An additional library, ArNetworking is used to write the server and client programs, but ArNetworking is not covered here. All ARIA programs presented in this text assume that the user is controlling their robot from its own on-board PC.

- **The Player Server**
  Player is a single device server that runs on the robot PC, providing control over the sensors and actuators [1]. It is language independent meaning that client control software can be written in any language that can open and control a TCP socket. Client-side libraries are currently available for C, C++, Tcl, LISP, Java, and Python. This guide covers use of the Player C++ library to provide high level client control programs for real Pioneer 3 robots and virtual ones created through Player's 2D simulator, Stage. The Player/Stage project is open source software and can be used to control many different robot makes, models and devices, for

example ER1 robots, kheperas and clodbusters, see Table 1.1 for a full list of supported robots. Player is architecturally independent and client programs can run both from the robot PC and from a remote networked PC, with no modification.

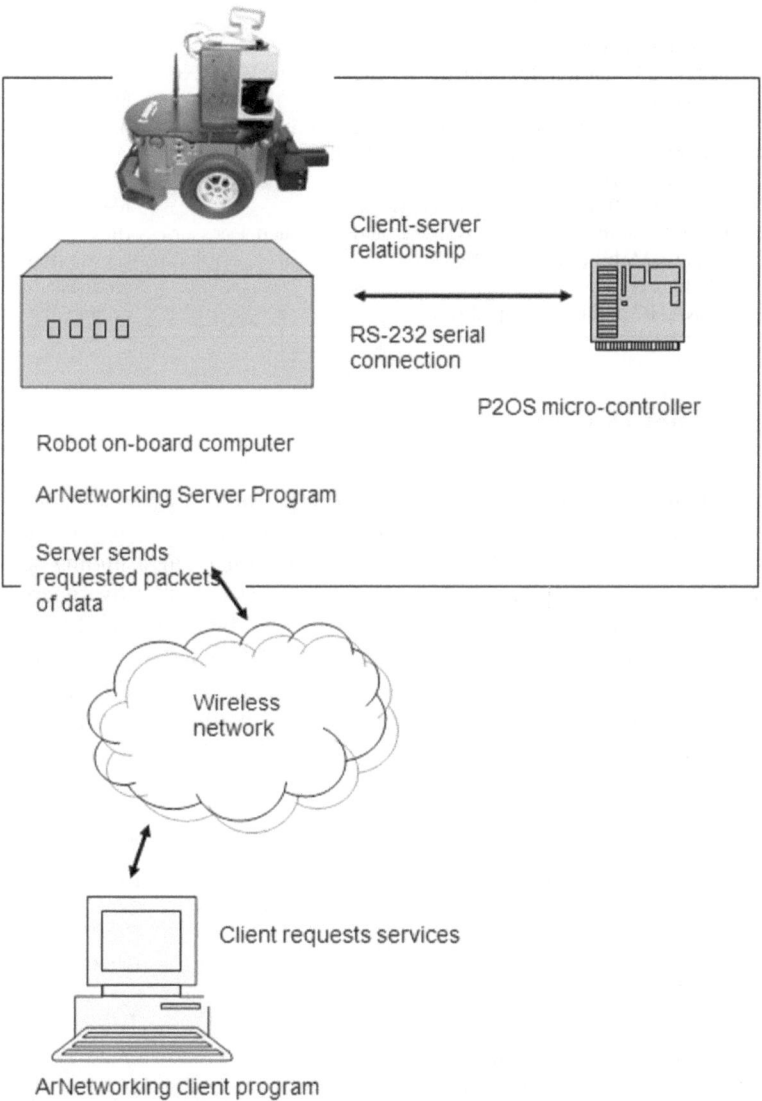

Fig. 1.2 Controlling the robot remotely using ArNetworking

Table 1.1 Robots supported by the Player server

Manufacturer	Device(s)	Driver
Acroname	Garcia	garcia
Botrics	Obot d100	obot
Evolution Robotics	ER1 and ERSDK robots	er1
iRobot	Roomba vacuuming robot roomba	roomba
K-Team	Robotics Extension Board (REB) attached to Kameleon 376BC	reb
	Khepera	khepera
MobileRobots	PSOS/P2OS/AROS-based robots (Pioneer, AmigoBot) and integrated accessories, including a CMUcam connected to the AUX port	p2os
Nomadics	NOMAD200 (and related) mobile robots	nomad
RWI/iRobot	RFLEX-based robots B21r, ATRV Jr and integrated accessories	rflex
Segway	Robotic Mobility Platform (RMP)	segwayrmp
Upenn Grasp	Clodbuster	clodbuster
Videre Design	ERRATIC mobile robot platform	erratic

1.3 Pioneer Robot Devices

The devices available for Pioneer 3 robots are listed in Table 1.2 and illustrated in Figures 1.3 and 1.4 below, which show two different hardware configurations. The devices include a SICK LMS200 laser range finder, sonar range finding sensors, a Canon VC-C4 pan-tilt-zoom camera, grippers, bumpers, and a five-degree-of-freedom (5D) arm. Table 1.2 shows which devices may be controlled by each of the programming interfaces, ARIA and Player, and considers both real and simulated robots. (MobileSim is the 2D simulator for ARIA and Stage is the 2D simualtor for Player.) ACTS is a blob finding software package compatible with ARIA, see Section 3.1, but Player uses a simpler proxy to support blob finding, see Chapter 5.

Table 1.2 Common devices available for the Pioneer 3 robot and software that supports them

Device	Description	ARIA (Real robot)	MobileSim 0.5.0	Player 2.0.5 (Real robot)	Stage 2.0.4
Sick LMS200 Laser (front 180°)	Range finding laser sensor	√	√	√	√
Pioneer Sonar Ring (8 front, 8 rear)	Range finding sonar sensors	√	√	√	√
Canon VC-C4 ptz camera	Pan-tilt-zoom camera	√		√	√
Simple blob finding device	For tracking colour			√	√
ACTS blob finder	For tracking colour	√		√	
Pioneer bumper pads (5 rear)	Collision detection	√	√	√	√
Pioneer 2D gripper	2 degrees freedom	√		√	√
Pioneer 5D arm	5 degrees freedom	√		√	

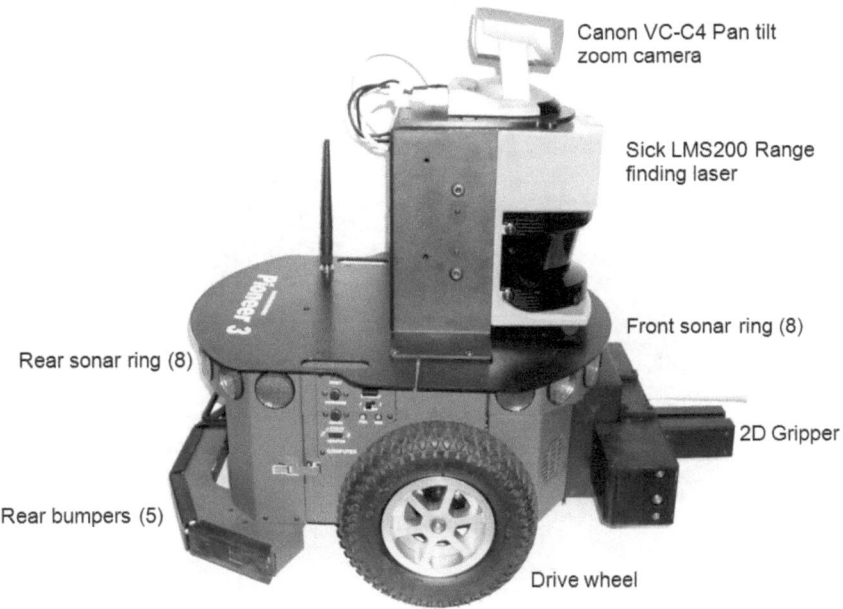

Canon VC-C4 Pan tilt
zoom camera

Sick LMS200 Range
finding laser

Front sonar ring (8)

Rear sonar ring (8)

2D Gripper

Rear bumpers (5)

Drive wheel

Fig. 1.3 Pioneer P3-DX sensors and actuators

It is clear from Table 1.2 that both ARIA and Player are fully able to support control of real Pioneer robots and their devices, but that the MobileSim simulator is much more limited than Stage in terms of the devices that can be simulated. Stage 2.0.4 is fully able to support all of the devices listed except for ACTS and for the 5D arm (which can only be simulated in 3D).

The next section is concerned with downloading and installing the necessary packages to run the ARIA API and the other MobileRobots resources.

1.4 MobileRobots Software Installations

Root permissions are required to install all of the MobileRobots software and any other programs (e.g. rpm or alien) that may be needed as part of the installation process. ARIA and its associated programs can be downloaded from the MobileRobots web site "http://www.mobilerobots.com", but a valid user ID and password must be supplied. Follow the "support and downloads" link and then the "software and firmware" link, which leads to a list of the available programs.

5D arm

Rear sonar ring (8)

Front sonar ring (8)

Rear bumpers (5)

Canon VC-C4 Pan tilt
zoom camera

Drive wheel

Fig. 1.4 Pioneer P3-DX sensors and actuators, alternative configuration

1.4.1 ARIA

The software must be installed both on the robot and on any PC where you intend
to compile, test or execute your control programs. At the time of writing ARIA is
on version 2.7.0 and this available for both Debian and Red Hat Linux. Download
the appropriate installation file and save it in /usr/src. If you have Red Hat compatible Linux and the Red Hat package manager installed on your system you can
install by using:

```
rpm -i ARIA-2.7-0-0.i386.rpm.
```

If you have Debian Linux then use:

```
dpkg -i libaria_2.7.0_i386.deb.
```

The software installs in the directory /usr/local/Aria.

1.4.2 Mapper3Basic

Mapper3Basic is presently at version 2.2.5 and is available for both Debian and Red Hat Linux. Download the appropriate installation file ("Mapper3Basic-2.2-5-0.i386.rpm" for Red Hat Linux or "mapper3-basic_2.2.5_i386.deb" for Debian) and save it in /usr/src. You can then install it in the same way as ARIA.

The software installs in the directory /usr/local/Mapper3Basic and the binary is /usr/local/bin/Mapper3Basic.

1.4.3 MobileSim

MobileSim is currently available for Red Hat and Debian Linux and is on version 0.5.0 at the time of writing. Download the appropriate Linux installation file ("MobileSim-0.5.0-0.i386.rpm" for Red Hat or "mobilesim_0.5.0_i386.deb" for Debian) and save it in /usr/src. Install using the rpm command or dpkg command.

The software installs in the directory /usr/local/MobileSim and the binary is /usr/local/bin/MobileSim.

1.4.4 ACTS

ACTS is currently only available for Red Hat Linux. Download the Red Hat Linux installation file, e.g. "ACTS-2.2-0.i386.rpm" and save it in /usr/src. If you have Red Hat Linux and the Red Hat package manager installed on your system you can install by using:

```
rpm -i ACTS-2.2-0.i386.rpm.
```

If you do not have Red Hat Linux you can still install the Red Hat package manager, for example, with Debian you can use apt-get install rpm and then install as above. However, there are often problems when trying to install source code intended for a different distribution of Linux. For example the install may indicate failed dependencies that are actually present; a common example is reporting that /bin/sh is required but not present. This sort of error can sometimes be overcome by using the --nodeps option, i. e. rpm -i --nodeps ACTS-2.2-0.i386.rpm, but this can often mean that the software fails to install properly as it does not check for required files. The best solution to this problem is to install alien on your machine. You can then convert the Red Hat installation file to your own distribution. For example, with Debian, use:

```
apt-get update
```

```
apt-get install alien
```

and then convert using:

```
alien --to-deb ACTS-2.2-0.i386.rpm.
```

In the above example a new file "acts-2.2-0.i386.deb" would be created. You can then install in the usual way, for example:

```
dpkg -i aria-2.2-1.i386.deb
```

for Debian. The software installs in the directory /usr/local/acts and the executable is the file /usr/local/acts/bin/acts.bin. After installation you will need to edit the file /etc/ld.so.config. This file is a list of paths to libraries, so you will need to add the following path to the list:

```
/usr/local/acts/lib.
```

You should then save the edited file and run ldconfig to update the libraries. When you try to run the binary you may get the following error message "error while loading shared libraries: libpng.so.2: cannot open shared object file: No such file or directory". This problem can be solved by creating a symbolic link to the file libpng.so. If you have this installed it will be in the directory /usr/lib. Go into that directory and use the following command: ln -s libpng.so libpng.so.2. Then try running the binary again.

1.5 Player and Stage Installations

The Player and Stage source programs are available for free download at:

"http://sourceforge.net/project/showfiles.php?group_id=42445".

This book consistently refers to the installation and use of Player and Stage version 2.0 of which the latest versions were 2.0.5 and 2.0.4 respectively at the time of writing. Although newer versions (2.1) are available, the programs written and used for demonstration in this book were all tested with version 2.0. However, they should work with version 2.1 with no problems.

The best sources of information for Player and Stage are the current online manuals [1] and [2] respectively. Other useful resources include the project authors' papers [4], [5] and [6] and the Player / Stage users searchable mailing list archive [7], which can often provide help with specific problems.

The default location for installation is /usr/local so you will need root permissions to install the software there. However, Player and Stage both allow instal-

lation in other folders so you can install them within your home directory if you do
not have root permissions on your machine, see Section 1.5.3 and Section 1.5.6.

1.5.1 Prerequisites

For Linux you should be running either a 2.4 or 2.6 kernel. You should also have
a recent version of gcc (the GNU Compiler Collection), preferably version 3.2 or
above. (Use gcc -v to check which version you have.) The following tools are
also required:

- gcc with C++ support (g++)
- autoconf
- automake
- libtool
- make

Player/Stage also depends on some third-party libraries. The main one (needed for
PlayerViewer and Stage) is the GIMP toolkit (GTK), which comes as standard with
most Linux distributions. If you do not have it you can install by using, for example:

```
apt-get install libgtk-dev
apt-get install libgtk2.0
apt-get install libgtk2.0-dev
```

on a Debian system. In addition, the GTK has some dependencies of its own. If
you want to run Stage you should install Player first.

1.5.2 Player - Default Location

If you want to run client programs on a real robot but from a remote machine you
will need to install Player on both the robot and the remote PC. The default location for the binary is /usr/local, so you will need root permissions if you want
to install it there. First download the source tarball, e.g. "player-2.0.5.tar.bz2" and
place it in the file /usr/src, then uncompress it using:

```
bzip2 -dc player-2.0.5.tar.bz2 | tar xf -.
```

After uncompressing a new directory in /usr/src (player-2.0.5) will have been
created . Go into this directory and configure Player by typing: ./configure. The
screen output from the configuration will show useful information such as which
drivers will be built. You can also check the file "config.log" for detailed information

about which tests failed and why. When you are satisfied with the configuration output compile by typing make and finally install Player by typing make install. On installation the Player binary is placed in /usr/local/bin, and the libraries (e.g., "libplayercore", "libplayerdrivers") will be placed in /usr/local/lib. Note that PlayerViewer, the GUI visualisation software for Player, should also be installed as part of the process and also resides in /usr/local/bin. To check that the software has installed properly you can type:

```
pkg-config --libs playercore
```

and the following libraries should be displayed on screen:

```
-lplayercore -lltdl -lpthread -lplayererror.
```

If this message is displayed instead: "Package playercore was not found in the pkg-config search path. Perhaps you should add the directory containing 'player-core.pc' to the PKG_CONFIG_PATH environment variable. No package 'player-core' found" then you will need to add the path /usr/local/lib/pkgconfig to your PKG_CONFIG_PATH. This can be done by editing your bashrc file (a hidden file in your home directory), i.e., by adding the following line to it:

```
export PKG_CONFIG_PATH=/usr/local/lib/pkgconfig:
```

```
$PKG_CONFIG_PATH.
```

Note that librtk2, the library for the robot toolkit that was an integral part of installation for older versions of Player and Stage has been deprecated and is not required for installation of Player 2.0 or Stage 2.0. The robot toolkit is now a part of Player.

1.5.3 Player - Selected Location

You can change the installation directory from the default by using the --prefix option and specifying the absolute path of the desired location when configuring. For example, if you need to install Player in your home directory (e.g. /home/amw/) because you do not have root access, use:

```
./configure --prefix=/home/amw/local/player.
```

Then run make and make install as before.

After installation the executables are in /home/amw/local/player/bin, and the libraries in /home/amw/local/player/lib. Various environment variables will also need to be set including the PATH, LD_LIBRARY_PATH and PKG_CONFIG_PATH. In the example above you would need to add the following to your .bashrc file.

```
export PKG_CONFIG_PATH=$PKG_CONFIG_PATH:home/amw/
local/player/lib/pkgconfig::$PKG_CONFIG_PATH
export PATH=/home/amw/local/player/bin:$PATH
export CPATH=/home/amw/local/player/include:$CPATH
export LD_LIBRARY_PATH=/home/amw/local/player/lib:
$LD_LIBRARY_PATH
```

The first line tells Stage where Player is. The second line shows where the Player binary is, so that you can run it by just typing player <configfile> instead of the full path name. The third line shows where the header files are for compiling Player programs and the fourth shows where the libraries are for linking.

1.5.4 Selecting Drivers

Certain drivers will be built and others omitted by default. To override the defaults you can use the --enable and --disable options when configuring, to enable and disable the compilation of certain drivers respectively. For example to enable the ACTS driver you would use:

```
./configure --enable-acts.
```

1.5.5 Stage - Default Location

You should first install Player and check that it is working correctly, see Section 1.5.2 and Section 1.5.3. The default location for stage is /usr/local. To install it here you will need root permissions. First download the source tarball, e.g. "stage-2.0.4.tar.gz" and place it in the folder /usr/src, then uncompress it using:

```
tar xzvfp stage-2.0.4.tar.gz.
```

After uncompressing, a new directory in /usr/src (stage-2.0.4) will have been created. Go into this directory and configure Stage by typing: ./configure. When you are satisfied with the configuration output, compile by typing make and finally install Stage by typing make install. Note that version 2.0 of Stage is not a program that runs as a standalone, i.e. there is no Stage binary. It merely provides a "plugin" for Player, which adds simulated robots; you run Player's binary with an appropriate configuration file, see Section 6.3. You can verify that the installation was successful by typing the following:

```
/usr/local/bin/player /usr/src/stage-2.0.4/
              worlds/simple.cfg.
```

You can just type `player simple.cfg` if you are already in the `worlds` directory and if your system knows where to find the Player binary.

1.5.6 Stage - Selected Location

You can change the installation directory from the default by using the `--prefix` option and specifying the absolute path of the desired location when configuring. For example, if you need to install Stage in your home directory (e.g. `/home/amw/`) because you do not have root access, use for example:

```
./configure --prefix=/home/amw/local/player.
```

Then run `make` and `make install` as before. It is important to use the same prefix that you used when installing Player as Stage is a plugin for Player.

The next chapter describes the process of constructing object-oriented control programs using the ARIA API. You will quickly learn how to connect to your robot, instantiate, add and connect to devices, control each sensor and actuator and integrate all of this into a single control program.

Chapter 2
Programming with the ARIA API

2.1 Getting Started

The best source of information is the online help document that comes with the software installation [14]. It is located in /usr/local/Aria and has the name "Aria-Reference.html". All the classes that form the ARIA library are listed and their attributes and methods are described there.

2.1.1 Compiling Programs

ARIA programs are compiled under Linux by using g++ on the command line. All programs must be linked to the ARIA library "lAria" and the additional libraries "lpthread" and "ldl". The ARIA library is located in /usr/local/Aria/lib and the header files are located in /usr/local/Aria/include. You will need to add the path /usr/local/Aria/lib to the file /etc/ld.so.conf and run ldconfig in order to access the libraries. As an example, suppose you have a control program named "test.cpp" and you wish to create a binary called "test". From the directory where "test.cpp" is located, you would type the following:

```
g++ -Wall -o test -lAria -ldl -lpthread -L/usr/local/
    Aria/lib -I/usr/local/Aria/include test.cpp.
```

Alternatively, a suitable bash script such as the example given below can be written to save typing:

```
#!/bin/sh

# Short script to compile an ARIA client
# Requires 2 arguments, (1) name of binary
```

```
# and (2) name of program to compile

if [ $# != 2 ]; then
    echo Require 2 arguments
    exit 1
fi


g++ -Wall -o $1 -lAria -ldl -lpthread -L/usr/local/Aria
/lib -I/usr/local/Aria/include $2
```

2.1.2 Connecting to a Robot

A method for sending and receiving data to and from the server must be specified. For real robots the server software for low level control runs on the micro-controller and communication between this and the robot PC is through a serial port. If you want to test your programs on a simulator first (on a remote PC) and then run them on a real Pioneer without changing the program, the best way to connect is to use the ArSimpleConnector and ArArgumentParser classes. The ArSimpleConnector class first tries to connect to a simulator if one is detected, otherwise it connects through the serial port of the real robot. For this to work you need to run the control program on the robot PC itself, i.e. connect to the robot first using ssh and then run the program. Unfortunately this involves copying the control program from the remote PC to the robot and recompiling. If you want to run the program directly from a remote PC you need to use the separate ArNetworking C++ library to create a server program that runs on the robot PC and a client program that runs on the remote PC. The server program sets up the services that the client program can then request. This involves writing a new control program and is beyond the scope of this guide, which assumes that you will run your program on the robot PC.

Below is an extract of a program that shows how to connect to a robot using Ar-SimpleConnector.

```
/* Include files */

#include "Aria.h"                                                     1
#include <stdio.h>                                                    2

/* Main method */

int main(int argc, char **argv)                                       3
{

/* The robot and its devices */

Aria::init();                              //Initialise ARIA library  4
```

```
ArRobot robot;                          //Instantiate robot              5

ArArgumentParser parser(&argc, argv);   //Instantiate argument parser    6
ArSimpleConnector connector(& parser);  //Instantiate connector          7

/* Connection to robot */

parser.loadDefaultArguments();          //Load default values            8

if (!connector.parseArgs())             //Parse connector arguments      9
   {
     cout << "Unknown settings\n";       //Exit for errors               10
     Aria::exit(0);                                                       11
     exit(1);                                                            12
   }

if (!connector.connectRobot(&robot))    //Connect to the robot          13
   {
     cout << "Unable to connect\n";      //Exit for errors              14
     Aria::exit(0);                                                     15
     exit(1);                                                           16
   }

robot.runAsync(true);                   //Run in asynchronous mode      17

robot.lock();                           //Lock robot during set up      18
robot.comInt(ArCommands::ENABLE, 1);    //Turn on the motors            19
robot.unlock();                         //Unlock the robot              20

Aria::exit(0);                          //Exit Aria                     21

}                                       //End main
```

"Aria.h" must be included with all programs (line 1) and before the ARIA library can be used it must be initialised by using Aria::init() (line 4). The ArRobot class (instantiated here in line 5) is the base class for creating robot objects that you can then connect devices to. An instance of the class essentially represents the base of a robot with no sensors attached and only the motors for actuators [12]. However, MobileRobots describe the class as the "heart" of ARIA as it also functions as the client-server gateway, constructing and decoding packets and synchronising their exchange with the micro-controller [14]. Standard server information packets (SIPs) get sent by the server to the client every 100 milliseconds by default. The ArRobot class runs a loop (either in the current thread by using the ArRobot::run() method or in a background thread by using ArRobot::runAsync()), which is synchronised to the data updates sent from the robot micro-controller. In the above program the ArRobot::runAsync() method is used (line 17) after connection has been established. Running the robot asynchronously like this ensures that if the connection is lost the robot will stop.

An ArArgumentParser object is instantiated here in line 6. This is a standard argument parser for maintaining uniformity between ARIA-based programs. It ensures that all the configurable elements of an ARIA program (robot IP address etc.)

are passed to it in the same way [12]. The constructor for ArSimpleConnector takes
a pointer to the ArArgumentParser object (line 7). The loadDefaultArguments()
method of ArArgumentParser is called in line 8. This allocates the default arguments required to connect to a local host (either MobileSim, see Section 3.2 or the
real robot). Once the default arguments are loaded they can be parsed to the ArSimpleConnector object by using its parseArgs() method (line 9). The connectRobot()
method can then be used to make the actual connection. A pointer to the ArRobot
object must be supplied as the argument (line 13).

Before running any commands the motors should be placed in an enabled state,
(line 19). It is advisable to lock the robot (line 18) to ensure that the command is
not interfered with by other users, and the robot should be unlocked afterwards (line
20). When the program ends ARIA must be exited using the syntax in line 21. If you
get a segmentation fault when running the program it may be necessary to remake
the files in /usr/local/Aria after installation.

2.2 Instantiating and Adding Devices

In ARIA devices fall into two categories, ranged devices (sonar, laser and bumpers),
which inherit from the ArRangeDevice class and non-ranged devices, (anything
else, e.g. a pan-tilt-zoom camera or a 2D gripper). There are differences in how
these types of device are associated with a robot.

2.2.1 Ranged Devices

Ranged devices are instantiated and then added to the robot using ArRobot's ad-
dRangeDevice() method, which takes a pointer to the device as its argument. Below
are some extracts of programs that show how to instantiate a sonar device, a laser
device and a set of bumpers, and also how to add them to an ArRobot object called
"robot".

```
ArRobot robot;                      //Instantiate the robot
ArSick laser;                       //Instantiate its laser
ArSonarDevice sonar;                //Instantiate its sonar
ArBumpers bumpers;                  //Instantiate its bumpers

robot.addRangeDevice(&sonar);       //Add sonar to robot
robot.addRangeDevice(&laser);       //Add laser to robot
robot.addRangeDevice(&bumpers);     //Add bumpers to robot
```

The laser device requires additional initialisation to other devices as it inherits
from the ArRangeDeviceThreaded class (which inherits from the ArRangeDevice
class). This means that it is a ranged device that can run in its own thread. It there-

fore requires additional connection to the robot using ArSimpleConnector's connectLaser() method, see line 8 of the program extract below.

```
/* Connection to laser */

Aria::init();                         //Initialise ARIA library    1
ArRobot robot;                        //Instantiate robot          2
ArSick laser;                         //Instantiate laser          3
robot.addRangeDevice(&laser);         //Add laser                  4
ArArgumentParser parser(&argc, argv); //Instantiate argument parser 5
ArSimpleConnector connector(& parser); //Instantiate connector      6

      .
      .
      .                               //Connect to robot
      .

laser.runAsync();                     //Asynchronous laser mode    7

if (!connector.connectLaser(&laser))  //Connect laser to robot     8
   {
      cout << "Can't connect to laser\n"; //Exit if error           9
      Aria::exit(0);                                               10
      exit(1);                                                     11
   }

laser.asyncConnect();                 //Asynchronous laser mode    12
```

Lines 1 to 6 instantiate the various objects and lines 8 to 11 make and check the connection. Asynchronous connection is specified in lines 7 and 12 and ensures that the laser will stop if the connection fails. An alternative way of connecting to the laser is shown below.

```
connector.setupLaser(&laser);

laser.runAsync();

if (!laser.blockingConnect())
   {
      cout << "Could not connect to SICK laser... exiting\n");
      Aria::exit(0);
      exit(1);
   }
```

2.2.2 Non-ranged Devices

Non-ranged devices do not inherit from ArRangeDevice so are not associated with
the ArRobot object in the same way. In fact, non-ranged devices may inherit from
other base classes, for example an ArVCC4 object (Canon VC-C4 pan-tilt-zoom
camera) inherits from the ArPTZ class. In general, the robot is added to non-ranged
devices instead of their being added to the robot. Sometimes this may be done as
part of the initialisation, for example the program extract below shows how a 2D
gripper and Canon VC-C4 pan-tilt-zoom camera are associated with the robot at the
same time as they are instantiated:

```
ArGripper gripper(&robot);    //Instantiate gripper and add robot
ArVCC4 ptz(&robot);           //Instantiate Canon VCC4 camera and add robot
```

On the other hand, the robot is added to a 5D arm object by first instantiating
the arm and then using its setRobot() method to add the robot, see Section 2.3.5 for
further details.

```
ArP2Arm arm;                  //Instantiate a 5D arm
arm.setRobot(&robot);         //Add robot to arm
```

An ACTS object (virtual blob finding device) uses its openPort() method both to
add the robot and to set up communication with the ACTS server running on the
robot, see Section 3.1 for further details.

```
ArACTS_1_2 acts;              //Instantiate an ACTS object
acts.openPort(&robot);        //Add robot and set up communication
                              //with ACTS server running on that robot
```

2.3 Reading and Controlling the Devices

Once devices have been instantiated and added to the robot, they can be controlled.
The rest of this chapter shows how this is achieved in ARIA for the Pioneer's motors,
sonars, laser, bumpers, 5D arm, 2D gripper and camera. Programming of the ACTS
blob finder is dealt with in Section 3.1.

2.3.1 The Motors

Motion commands can be issued explicitly by using the setVel(), setVel2() and
setRotVel() methods of the ArRobot class; the setVel() method sets the desired translational velocity of the robot in millimetres per second, setVel2() sets the velocity
of the wheels independently and setRotVel() sets the rotational velocity of the robot

in degrees per second. In addition there are the setHeading() and setDeltaHeading() methods, which change the robot's absolute and relative orientation (in degrees) respectively. There is also a method to move a prescribed distance (move()) and a method for stopping motion (stop()). If a positive double is supplied as the argument for move(), the robot moves forwards. If a negative double is supplied the robot moves backwards. Some examples of these methods are shown below. All these use a previously declared ArRobot object called "robot".

```
robot.setVel(200);           //Set translational velocity to 200 mm/s
robot.setRotVel(20);         //Set rotational velocity to 20 degrees/s
robot.setVel2(200,250);      //Set left wheel speed at 200 mm/s
                             //Set right wheel speed at 250 mm/s
robot.setHeading(30);        //30 degrees relative to start position
robot.setDeltaHeading(60);   //60 degrees relative to current orientation
robot.move(200);             //Move 200 mm forwards
```

Other methods of interest are setAbsoluteMaxTransVel() and getAbsoluteMax-TransVel(), which set and get the robot's maximum allowed translational speed. This is useful if you do not want your robot to exceed a given speed for safety reasons. The methods setAbsoluteMaxRotVel() and getAbsoluteMaxRotVel() do the same for rotational speed and the methods getVel() and getRotVel() return the robot's translational and rotational speeds respectively, as double values.

Note that more complex forms of motion can be achieved by creating action classes that inherit from ARIA's ArAction class and adding the actions to the robot. The actions then provide motion requests that can be evaluated and combined to produce a final desired motion. In this way complex behaviours can be achieved. However you can create actions that do not inherit from ArAction if you do not want to implement this particular behaviour architecture. Further details about ArActions are provided in Chapter 4. The program below shows user-written methods "wander()" and "obstacleAvoid()" that implement simple wandering and obstacle avoidance behaviours respectively. These methods do not inherit from ArAction.

```
/*
*---------------------------------------------------------------
* Wandering mode
*---------------------------------------------------------------
*/

void wander(double speed, ArRobot *thisRobot)
{

int rand1;                           //Whether to change direction
int rand2;                           //Used to decide angle of turn
int rand3;                           //Used to decide direction of turn
int dir;                             //Direction of turn

srand(static_cast<unsigned>(time(0))); //Set seed
```

```
rand1 = (rand()%2);                  //Get random no. between 0 and 1

if (rand1 == 0)                      //1 in 2 chance of turning
  {
    rand2 = (rand()%10);             //Get random no. between 0 and 9
    rand3 = (rand()%2);              //Get random no. between 0 and 1

    switch(rand3)                    //Get direction based on rand3
      {
        case 0:dir = -1;break;       //Turn right
        case 1:dir = 1;break;        //Turn left
      }
  }else
  {
    dir = 0;                         //Don't turn
    rand2 = 0;
  }

thisRobot->setRotVel(rand2*10*dir/2); //Set rotational speed
thisRobot->setVel(speed);            //Set translational speed
}

/*
*-------------------------------------------------------------
* Obstacle avoidance mode
*-------------------------------------------------------------
*/

void obstacleAvoid(double minAng, double driveSpeed, ArRobot *thisRobot)
{

double avoidAngle;                   //Angle to turn to avoid obstacle

if (minAng >= 0 && minAng < 46 )     //If obstacle is to the left
  {
    cout << "TURNING RIGHT!\n";
    avoidAngle = -30.0;              //Turn right
  }

if (minAng > -46 && < 0)             //If obstacle is to the right
  {
    cout << "TURNING LEFT!\n";
    avoidAngle = 30.0;               //Turn left
  }

thisRobot.setRotVel(avoidAngle);     //Set rotational speed
thisRobot.setVel(driveSpeed);        //Set translational speed

}
```

2.3.2 *The Sonar Sensors*

Sonar devices are instantiated and added to the robot as described in Section 2.2.1. To obtain the closest current sonar reading within a specified polar region, the currentReadingPolar() method of the ArRangeDevice class can be called. The polar region is specified by the startAngle and endAngle attributes (in degrees). This goes counterclockwise (negative degrees to positive). For example if you want the slice between -45 and 45 degrees, you must enter it as -45, 45. Figure 2.1 below shows the angular positions ARIA assigns to each of the sonar on the Pioneer robots. The closest reading is returned by the method, but is the distance from the object to the assumed centre of the robot. To obtain the absolute distance the robot radius should be subtracted. This can be done by calling ArRobot's getRobotRadius() method. The angle at which the closest reading was taken is obtained by supplying a pointer to the double variable holding that value. An example program that implements the currentReadingPolar() method is shown below:

```
ArRobot robot;                  //Instantiate the robot
ArSonarDevice sonar;            //Instantiate its sonar
robot.addRangeDevice(&sonar);   //Add sonar to robot
        .
        .                       //Connect to robot
        .

double reading, readingAngle;   //To hold minimum reading and angle
reading = sonar.currentReadingPolar(-45,45,&readingAngle);
                                //Get minimum reading and angle
```

If raw sonar readings are required then the getSonarReading() method of the ArRobot class can be called. The index number of the particular sonar is used as the argument. The method returns a pointer to an ArSensorReading object. By calling the getRange() and getSensorTh() methods of this class you can obtain both the reading and its angle. If you need all the sonar readings then you should first determine the number of sonar present using the getNumSonar() method of the ArRobot class and then call the getSonarReading() method in a loop. An example user-written method "getSonar()", which prints all the raw sonar readings and their angles is shown below:

```
/*
*------------------------------------------------------------
* Print raw sonar data
*------------------------------------------------------------
*/

void getSonar(ArRobot *thisRobot)
{
```

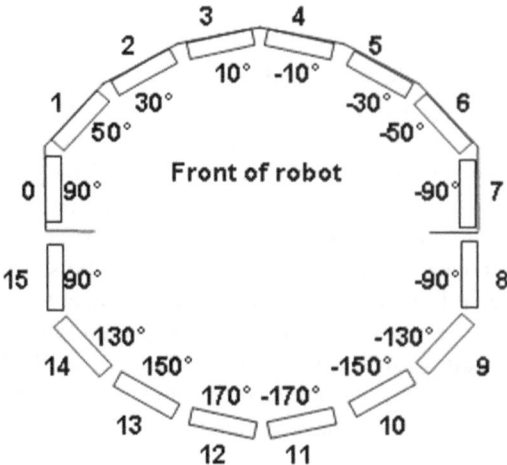

Fig. 2.1 The angular positions of the sonar sensors

```
int numSonar;                          //Number of sonar on the robot
int i;                                 //Counter for looping

numSonar = thisRobot->getNumSonar();  //Get number of sonar
ArSensorReading* sonarReading;         //To hold each reading

for (i = 0; i < numSonar; i++)         //Loop through sonar
  {
    sonarReading = thisRobot->getSonarReading(i);
                                       //Get each sonar reading
    cout << "Sonar reading " << i << " = " << sonarReading->getRange()
      << " Angle " << i << " = " <<
    sonarReading->getSensorTh() << "\n";
  }
}
```

The sonar can be simulated using MobileSim, see Section 3.2.

2.3.3 The Laser Sensor

Laser devices are instantiated, added to the robot and connected as described in section 2.2.1. As both the sonar and laser devices inherit from the ArRangeDevice class, the currentReadingPolar() method can also be used with the laser, see Section 2.3.2. An example program is shown below:

```
ArRobot robot;                         //Instantiate the robot
```

```
ArSick laser;                    //Instantiate its laser
robot.addRangeDevice(&laser);    //Add laser to robot
        .
        .                        //Connect to robot
        .

double reading, readingAngle;    //To hold minimum reading and angle
reading = laser.currentReadingPolar(-45,45,&readingAngle);
                                 //Get minimum reading and angle
```

Another useful method to invoke is the checkRangeDevicesCurrentPolar() method of the ArRobot class. This checks all of the robot's ranged sensors in the specified range, returning the smallest value. An example using an ArRobot object called "robot" is shown below.

```
double reading = robot.checkRangeDevicesCurrentPolar(-45,45);
```

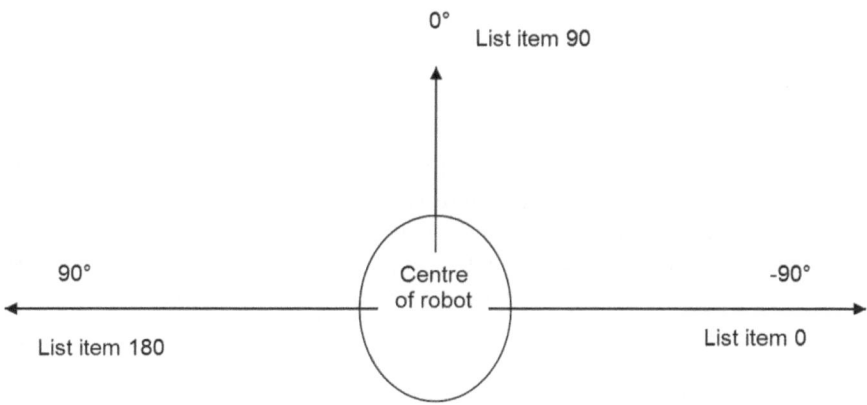

Fig. 2.2 Laser readings and their positions on the robot (181 readings)

If raw laser readings are required then the procedure is slightly more complex than for sonar sensors as it involves using lists. The method to call is the getRawReadings() method of the ArSick class. This returns a pointer to a list of ArSensorReading object pointers. You will need to loop through this list to obtain the values and angles, so you will also need to declare an iterator object for the list as well as the list itself. You can then loop through each ArSensorReading pointer and obtain its reading and angle by calling its getRange() and getSensorTh() methods. An example user-written method "getLaser()", which prints all the raw laser readings and their angles is shown below:

```
/*
*-----------------------------------------------------------
* Print raw laser data
*-----------------------------------------------------------
*/

void getLaser(ArSick *thisLaser)
{

/* Instantiate sensor reading list and iterator object */
const std::list<ArSensorReading *> *readingsList;
std::list<ArSensorReading *>::const_iterator it;
int i = -1;                       //Loop counter for readings

readingsList = thisLaser->getRawReadings();
                            //Get list of readings
                            //Loop through readings
for (it = readingsList->begin(); it != readingsList->end(); it++)
  {
    i++;
                            //Output distance and angle
    cout << "Laser reading " << i << " = " << (*it)->getRange()
        << " Angle " << i << " = " << (*it)->getSensorTh() << "\n";
  }
}
```

By default the laser should return 181 readings, see Figure 2.2 for the angular positions of each reading. If you require two readings for each degree then you should add the argument -laserincrement half when calling your control program. Further details about the SICK LMS200 laser and its operation can be found in [19]. Note that the laser can be simulated using MobileSim, see Section 3.2.

2.3.4 The Bumpers

Bumpers are instantiated and added to the robot as described in Section 2.2.1. Once bumpers have been declared you can obtain their state by calling the getStallValue() method of the ArRobot class. An example program using an ArRobot object called "robot" is shown below:

```
int rearBump=0;                      //State of bumpers and wheels
int numBumpers;                      //Number of bumpers

numBumpers = robot.getNumRearBumpers();   //Find number of bumpers
rearBump = robot.getStallValue();         //Get stall status
```

Table 2.1 below shows how to interpret the integer value returned by the getStallValue() method. First convert the integer to a binary number and store it in two bits.

For example if 6 was returned this would be 0000000000000110. The interpretation of the integer 6 is that rear bumpers 1 and 2 were bumped. On the Pioneers bumper 1 is the right most rear bumper and bumper 5 is the left most rear bumper, see Figure 2.3. If an integer value of 32 was returned this would mean that bumper 5 was bumped. However, if the left wheel was stalled the integer value would be 1. If the right wheel was stalled it would be 256 and if both were stalled it would be 257.

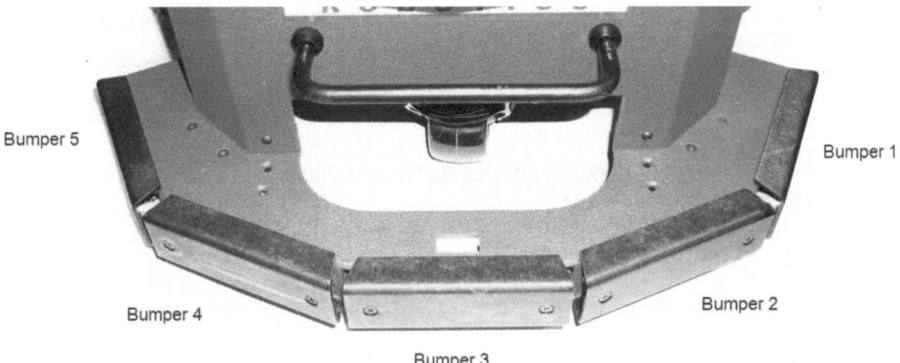

Fig. 2.3 Rear bumpers on the Pioneer robots

Table 2.1 Interpreting the stall integer

Example binary	0	0	0	0	0	0	1	1	0
Bumper number	Right wheel stall	-	-	5	4	3	2	1	Left wheel stall
Decimal component	256	128	64	32	16	8	4	2	1

If you need to check that the correct number of bumpers are being recognised (there are 5 rear bumpers on Pioneer P3-DX robots), then you can call the get-NumRearBumpers() method of the ArRobot class, which returns an integer value. There are also methods for checking the number of front bumpers, getNumFront-Bumpers(), and for checking whether front and rear bumpers are present, hasFront-Bumpers() and hasRearBumpers(), which both return boolean values. The program below shows the implementation of a user-written method "escapeTraps()", which uses the bumpers to determine where a bump has occurred and how to escape from it. The integer value "bumpVal" supplied to the method should be the result of calling ArRobot::getStallValue() . The double value "minRearReading" should be the smallest reading from the rear sonar, to determine whether the robot should reverse

out of the trap or move forwards. Notice that the method ArUtil::sleep() is called to
allow the robot time to carry out the motion commands; the argument to this is in
milliseconds.

```
/*
*--------------------------------------------------------------
* Escape traps mode
*--------------------------------------------------------------
*/

void escapeTraps(int bumpVal, double speed, double minRearReading,
    ArRobot *thisRobot)
{

if (bumpVal == 0)
  {
    cout << "TRAPPED AT FRONT MOVING BACKWARDS!\n";
    thisRobot->setRotVel(20);
    thisRobot->setVel(-1*speed);                       //Reverse
    ArUtil::sleep(2000);
  }else
  if (bumpVal > 1 && bumpVal < 63)                     //If any bumper registers
    {
      cout << "TRAPPED BEHIND MOVING FORWARDS!\n";
      thisRobot->setRotVel(20);
      thisRobot->setVel(speed);                        //Move forwards
      ArUtil::sleep(2000);
    }else
    if (bumpVal == 1 || bumpVal == 256 || bumpVal == 257)
                                                       //Either wheel has stalled
      {
        cout << "TRAPPED - MOVING EITHER FORWARD OR BACKWARDS!\n";

        if (minRearReading < 200)                      //Trapped at back
          {
            thisRobot->setRotVel(20);
            thisRobot->setVel(speed);                  //Move forwards
            ArUtil::sleep(2000);
            cout << "GOING FORWARDS TO ESCAPE\n";
          }else                                        //Not trapped at back
          {
            thisRobot->setRotVel(20);
            thisRobot->setVel(-1*speed);               //Move backwards
            ArUtil::sleep(2000);
            cout << "GOING BACKWARDS TO ESCAPE\n";
          }
      }
  }
}
```

Note that the MobileSim simulator does not generate bump signals other than the
right and left wheel stall signals, see Section 3.2.

2.3.5 The 5D Arm

ArP2Arm is the interface to the AROS/P2OS-based Pioneer arm servers. The arm is attached to the robot's micro-controller via an AUX serial port and the arm servers manage the serial communications with the arm controller [14]. The physical arm has 6 open-loop servo motors and 5 degrees of freedom, see [18] for more details. The end effector is a gripper with foam-lined fingers that can manipulate objects up to 150 g in weight. Table 2.2 lists the joints, which are illustrated in detail in Figure 2.4 and Figure 2.5.

Table 2.2 Joints list for the Pioneer 5D arm

Joint Number	Description
1	Rotating base
2	Pivoting shoulder
3	Pivoting elbow
4	Rotating wrist
5	Gripper mount (pivoting)
6	Gripper fingers

Fig. 2.4 5D arm gripper detail

Rotating wrist

Gripper mount
(pivoting)

Pivoting elbow

Gripper fingers

Pivoting shoulder

Rotating base

Fig. 2.5 The joints on the Pioneer

An ArP2Arm object is instantiated and associated with the robot as described
in Section 2.2.2, see also lines 1 to 3 of the program below. Note that the Ar-
Robot object that attaches to it must be run in its own thread, i.e. you should use
ArRobot::runAsync() if you are using the 5D arm. Following instantiation the arm
must be initialised first (line 4). This process communicates with the robot, checking
that an arm is present and in good condition [14]. The servos must also be powered
on (line 8) before the arm can be used. The program below shows how to do this:

```
ArRobot robot;                     //Instantiate a robot            1
ArP2Arm arm;                       //Instantiate a 5D arm           2
arm.setRobot(&robot);              //Add arm to robot               3

if (arm.init() != ArP2Arm::SUCCESS)   //Initialize the arm         4
  {
    printf("Arm initialization failed.\n");                        5
    Aria::exit(0);                                                 6
    exit(1);                                                       7
  }
arm.powerOn();                     //Turn on the arm                8
ArUtil::sleep(4000);               //Wait for arm to stop shaking   9
```

Note that the arm can shake for up to 2 seconds after powering on and if it is told to move before it stops shaking then it can shake even more violently. The powerOn() method of the ArP2Arm class waits 2 seconds by default but it is advisable to include an extra sleep statement as an added precaution (line 9).

The joints in the arm can be controlled by using the ArP2Arm::moveTo() method. This takes three arguments: an integer which specifies which joint is to be controlled, a float which specifies the position to move the joint to (in degrees) and an unsigned char, which specifies the speed of movement. If a velocity of 0 is specified then the current speed is used. Note that each joint has a -90 to 90 degree range approximately, but this can differ between designs. On the Pioneers all the joints rotate through at least 180 degrees, except the gripper fingers. The program below shows commands that move the arm joints (lines 1 to 5) and the fingers (line 6).

```
arm.moveTo(1,45,40);        //Set each joint              1
arm.moveTo(2,50,40);                                      2
arm.moveTo(3,20,40);                                      3
arm.moveTo(4,10,40);                                      4
arm.moveTo(5,15,40);                                      5
arm.moveTo(6,30,40);        //Set gripper                 6
ArUtil::sleep(6000);                                      7
arm.park();                 //Home arm and power it off   8
arm.uninit();               //Uninitialize the arm        9
```

The gripper at the end of the Pioneer arm is treated like the joints, where the angle passed is proportional to the amount of closing, i.e. to move it you just send the moveTo() command to joint 6. There is a public attribute ArP2Arm::NumJoints that allows the number of joints to be determined. By declaring a P2ArmJoint object it is also possible to obtain information about the state of that joint. This is done by using the getJoint() method of the ArP2Arm class and by reference to the myVel, myHome, myCenter, myPos, myMin, myMax and myTicksPer90 attributes of the class. The program extract below illustrates this:

```
ArRobot robot; ArP2Arm arm;
arm.setRobot(& robot);
if (arm.init() != ArP2Arm::SUCCESS)
  {
    cout << "Arm did not initialise\n";
    exit(1);
  }
P2ArmJoint *joint;
for (i=1, i<ArP2Arm::NumJoints; i++)
  {
    joint = arm.getJoint();
    cout << "Joint " << i << "velocity " << myVel << "home "
      << myHome << "\n";
  }
```

After use the arm should be set to its home position, powered off and unini-tialised. Lines 8 and 9 of the previous program show how this is achieved. The park() method both homes the arm and powers it off. This can also be accomplished with the separate methods home() and powerOff(). The home() method takes an in-teger value as its argument. If -1 is specified all joints are homed at a safe speed. If a single joint is specified only that joint is homed at the current speed. The powerOff() method should only be called when the arm is in a good position to power off as it will go limp. It is safer to use park() as this homes the arm first before it is powered off.

There are a number of other joint controlling methods that can be used. These include moveToTicks(), moveStep(), moveStepTicks(), moveVel() and stop(). The moveToTicks() method works in a similar way to the moveTo() method except the position is specified in ticks instead of degrees. A tick is the arbitrary position value that the arm controller uses. It uses a single unsigned byte to represent all the possi-ble positions in the range of the servo for each joint, so the range is 0 to 255 and this is mapped to the physical range of the servo. This is a lower level of arm control than using moveTo(). The moveStep() method also works in a similar way to moveTo() except that it moves a joint through the specified number of degrees rather than to a fixed position. The moveStepTo() method moves a joint through a specified number of ticks. The moveVel() method sets a particular joint to move at a specified velocity. It takes two integers, the first specifies the joint and the second specifies the veloc-ity. The desired velocity is actually achieved by varying the time between each tick movement. Thus, the attribute value supplied is actually the number of milliseconds to wait between each point in the path; 0 is the fastest, 255 is the slowest and a rea-sonable range is between 10 and 40. Calling the stop() method simply stops the arm from moving. This overrides all other actions except for initialisation. The 5D arm cannot be simulated using MobileSim as a 3D simulator is required for robot arms.

2.3.6 The 2D Gripper

The 2D gripper is instantiated and added to the robot as described in Section 2.2.2. Physically it is a two degree of freedom manipulation accessory that attaches to the front of the robot, see Figure 1.3 and Figure 2.6. It has paddles and a lift mechanism driven by reversible DC motors, with embedded limit switches that sense the paddle and lift positions. The paddles contain a grip sensor and front and rear infrared break beam switches that close horizontally until they grasp an object or close on themselves. Further details about the gripper device can be found in [14].

Table 2.3 shows the commands (i.e., all the methods of the ArGripper class) that can be used to determine the state of the gripper, the integer values that they return and how these are interpreted. Note that the getGripState() method returns a value of 2 (closed) both when the grippers are closed around an object and when they are fully closed on themselves. The integer value 0 (between open and closed) refers to their being in a moving state, not to their semi-closure. Note also that the paddles are always triggered when the gripper is closed. When the gripper is open they are triggered only when they are touched with sufficient pressure. Table 2.4 shows the ArGripper methods that can be used to control the gripper.

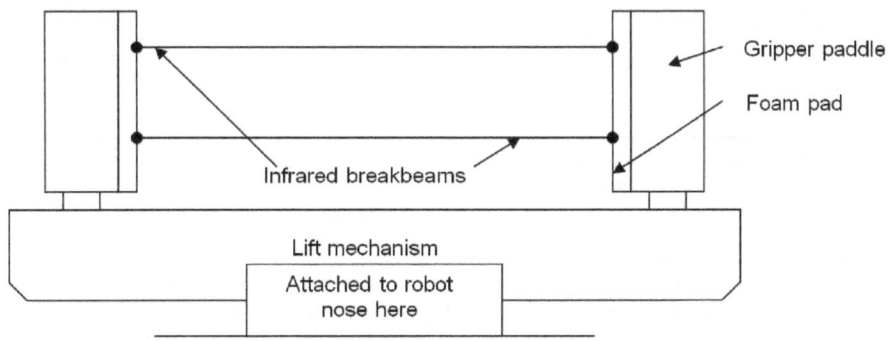

Fig. 2.6 The Pioneer 2D gripper

The program extract below shows use of the gripper's getType() method to check that a gripper is present before it is deployed for action.

```
ArRobot (robot);                          //Instantiate robot
ArGripper gripper(&robot);                //Instantiate gripper and add robot

int gripType;                             //Type of gripper
gripType = gripper.getType();             //Get gripper type
if (gripType != ArGripper::NOGRIPPER)     //If gripper is present
  {
    gripper.gripperDeploy();              //Open and raise gripper
    ArUtil::sleep(4000);                  //Wait while this completes
  }
```

The program sample below shows a user-written method to test the state of the break beams and the paddles and to close the grippers if they are broken by a can. After closure, the state of the break beams is tested again to make sure the can was successfully grabbed. The method returns a boolean value, which indicates whether the grab was successful or not.

```
/*
*----------------------------------------------------------------
* Can gripping routine
*----------------------------------------------------------------
*/

bool canGrip(ArGripper *thisGripper, ArRobot *thisRobot)
{
```

Table 2.3 Methods to obtain the gripper states

Method	Description	Returns	Interpretation
getGripState()	The collective state of the paddles	0 1 2	Between open and closed Open Closed
getPaddleState()	The individual state of the paddles	0 1 2 3	Not triggered Left triggered Right triggered Both triggered
getBreakBeamState()	The state of the break beams	0 1 2 3	None broken Inner beam broken Outer beam broken Both beams broken
getType()	Type of gripper The returned integer maps to an ARIA -defined enumeration value shown in brackets	0 1 2 3 4	Query type (QUERYTYPE) General input output (GENIO) User input output (USERIO) Packet requested from robot (GRIPPAC) No gripper present (NOGRIPPER)

Table 2.4 Methods to control the gripper

Command	Interpretation
gripOpen()	Opens the gripper paddles
gripClose()	Closes the gripper paddles
liftUp()	Raises the lift to the top
liftDown()	Lowers the lift to the bottom
gripperDeploy()	Puts the gripper in a deployed position, ready for use
gripStop()	Stops the gripper paddles
liftStop()	Stops the lift
gripperHalt()	Halts the lift and the gripper paddles

```
int beamState;                              //State of break beams
int paddleState;                            //State of the paddles
bool grippedCan = false;                    //Whether can gripped

beamState = thisGripper->getBreakBeamState();   //Get state of beams
paddleState = thisGripper->getPaddleState();    //Get paddle state

cout << "Gripper state is " << gripState << " \n";
cout << "Beam state is " << beamState << " \n";

/* If any beam is broken or paddles are triggered */
if (beamState == 1 || beamState == 2 || beamState == 3 ||
     paddleState == 1 || paddleState == 2 || paddleState == 3)
  {
```

```
    thisRobot->setVel(0);
    thisGripper->gripClose();                       //Grasp can
    ArUtil::sleep(4000);
    beamState = thisGripper->getBreakBeamState(); //Get beam state

    if (beamState == 1 || beamState == 2 || beamState == 3)
      {
        grippedCan = true;
        thisGripper->liftUp();                      //Lift can
        ArUtil::sleep(4000);
      }else
      {
        thisGripper->gripOpen();                    //Re-open as no can
        ArUtil::sleep(2000);
        grippedCan = false;
      }
  }

return grippedCan;                                  //Whether grab succeeded
}
```

The user-written method below releases a gripped can, reverses and then turns the robot. Note that the 2D gripper cannot be simulated using MobileSim, see Section 3.2.

```
/*
*-------------------------------------------------------------
* Can dropping routine
*-------------------------------------------------------------
*/

void canDrop(double speed, ArGripper *thisGripper, ArRobot *thisRobot)
{
thisRobot->setVel(0);                           //Stop moving
thisGripper->liftDown();                        //Lower gripper
ArUtil::sleep(4000);
thisGripper->gripOpen();                        //Open gripper
ArUtil::sleep(4000);
thisRobot->setVel(-1*speed);                    //Reverse slowly
ArUtil::sleep(4000);
thisRobot->setRotVel(90);                       //Turn away
ArUtil::sleep(2000);
}
```

2.3.7 The Pan-tilt-zoom Camera

A pan-tilt-zoom camera is instantiated and added to the robot as described in Section 2.2.2. Most Pioneers come with a Canon VC-C4 camera (see [21] and [20] for more details), so the ArVCC4 class, which inherits from the ArPTZ class, must be

used. Once instantiated the camera must first be initialised using the init() method
of the ArVCC4 class. The program extract below shows how to do this:

```
ArRobot(robot);                          //Instantiate robot
ArVCC4 ptz(&robot);                      //Instantiate ptz and add robot

bool ptzInitialized;                     //Whether ptz initialized

ptzInitialized = ptz.init();             //Initialize ptz
ArUtil::sleep(4000);
```

Once initialised, the camera can be controlled using the pan(), tilt() and zoom()
methods of the ArVCC4 class. These move the camera to a specified angle in degrees, which must be an integer value. In addition, the tiltRel() and panRel() methods can also be used to tilt or pan the camera relative to its present position. Other
useful methods include, panTiltRel() and panTilt() which perform the pan and tilt
operations together. Here, two integers representing the degrees of pan and tilt respectively are taken as arguments. The current angles of the camera can be obtained
by calling getPan(), getTilt() and getZoom(). The methods getMaxPosPan() and get-
MaxNegPan() retrieve the highest positive and lowest negative values that the camera can pan to (in degrees). The methods getMaxPosTilt() and getMaxNegTilt() do
the same for the tilt angles.

The user-written method below performs a simple camera movement exercise,
panning and tilting the camera through its full range continuously. Note that a pan-
tilt-zoom camera cannot be simulated using MobileSim, see Section 3.2.

```
/*
*-----------------------------------------------------------
* PTZ exercise mode
*-----------------------------------------------------------
*/

void ptzExercise(int inc, bool initPTZ, ArVCC4 *thisPTZ)
{

typedef enum                             //Tilt up or down
{
    up_U,
    down_D,
} VertDirection;

typedef enum                             //Pan left or right
{
    left_L,
    right_R,
} HorizDirection;

int panAngle;                            //Current pan angle
int tiltAngle;                           //Current tilt angle
int lowPan;                              //Lowest pan angle
int lowTilt;                             //Lowest tilt angle
```

```
int highPan;                              //Highest pan angle
int highTilt;                             //Highest tilt angle
HorizDirection hDir;                      //Horizontal direction
VertDirection vDir;                       //Vertical direction

if (initPTZ == true)                      //Camera initialization success
  {
    panAngle = thisPTZ->getPan();         //Get current pan
    tiltAngle = thisPTZ->getTilt();       //Get current tilt
    highPan = thisPTZ->getMaxPosPan();    //Get max pan
    lowPan = thisPTZ->getMaxNegPan();     //Get min pan
    highTilt = thisPTZ->getMaxPosTilt();  //Get max tilt
    lowTilt = thisPTZ->getMaxNegTilt();   //Get min tilt

    cout << "Pan = " << panAngle << " Tilt = " << tiltAngle << "\n";

    if (panAngle == highPan && tiltAngle == highTilt)
      {
        cout << "Changing direction to L and D\n";
        hDir = left_L;                    //Change directions
        vDir = down_D;
      }
    if (panAngle == lowPan && tiltAngle == lowTilt)
      {
        cout << "Changing direction to R and U\n";
        hDir = right_R;                   //Change directions
        vDir = up_U;
      }
    if (hDir == right_R)                  //If going right
      {
        thisPTZ->panRel(inc);             //Increment pan right
      }else
      {
        thisPTZ->panRel(-1*inc);          //Increment pan left
      }
    if (vDir == up_U)                     //If going up
      {
        thisPTZ->tiltRel(inc);            //Increment pan up
      }else
      {
        thisPTZ->tiltRel(-1*inc);         //Increment pan down
      }
  }else
  {
    cout << "Cannot initialize camera\n"; //Error message
  }
}
```

The next chapter looks at some of the other software packages offered by MobileRobots including ACTS, Mapper3Basic and the simulator MobileSim.

Chapter 3
Other MobileRobots Inc. Resources

3.1 ACTS Software

The ActivMedia Color Tracking System (ACTS) is a virtual vision sensor that allows tracking of coloured objects on 32 different channels. One of its components is a server program that can run on the robot PC. When running this server the user also supplies a configuration file that contains the locations of colour lookup tables (.lut files). These files hold the information about the colours that should be tracked and there is one file for each channel. The files are created by "training" each channel with a client program called EZ-train, using either still images or live video pictures. The best source of information about ACTS is the user guidebook, see [16].

In order to run ACTS as a blob finder device, a frame grabber, camera and supporting driver software are required. On the Pioneers the frame grabber is the device that Linux refers to as /dev/video0 and the camera is the Canon VC-C4. An X-window system is needed to train the channels using EZ-train, but Pioneer robots' on board PCs do not usually run X-windows. A way around this is to install the ACTS software both on the robot PC (for blob finding) and on the remote lab PC, which runs X-windows (for training the channels). It is also possible to export the graphical display over a network so that one machine runs and another displays the graphics, but the details of this are beyond the scope of this book.

3.1.1 Training the Channels

Once the software is installed on the remote PC and the robot PC run the binary on the remote PC by typing ./acts.bin from the directory /usr/local/acts/bin. Both the server and the EZ-train client will then run on the remote PC and two windows should be created, the image window and the control panel window for the EZ-train client. These are illustrated below in Figure 3.1 and Figure 3.2.

By default the image loaded into the image window is a still image of a Pioneer robot, however, you can specify a different image to load by typing `./acts.bin -f <imagename.ppm>`, where <imagename> is the name of the image you wish to load. Whilst running the program you can also change image by selecting *File →* *Load Image* from the menu and selecting your image. All images must be portable pixmap files with the extension .ppm, but it is usually very easy to save pictures in this format for example using KSnapshot or to convert them using graphical packages such as GIMP.

Fig. 3.1 EZ-train image window

Suppose you have a robot that has to pick up red cans in its gripper and you wish to train the ACTS system to track the red cans. By running the ACTS server on the robot (which runs another server program called savServer by default) and a client program called savClient on the remote PC, you can obtain live images from the robot camera on the remote PC. You can then capture some of the images of the cans as seen by the robot (using KSnapshot for example) and use these to train one of the channels to recognise the red cans. When you first run ACTS on the robot, you will need to specify a configuration file, but you will not have created one yet as you will not have trained any channels. You will therefore need to create a "dummy" configuration file in a simple text editor, which has the format illustrated below:

```
1 <NoChannel>
1 <NoChannel>
1 <NoChannel>
        .
        .
        .
        .
1 <NoChannel>
1 <NoChannel>
1 <NoChannel>
```

The file consists of "1 <NoChannel>" repeated 32 times on separate lines. Once

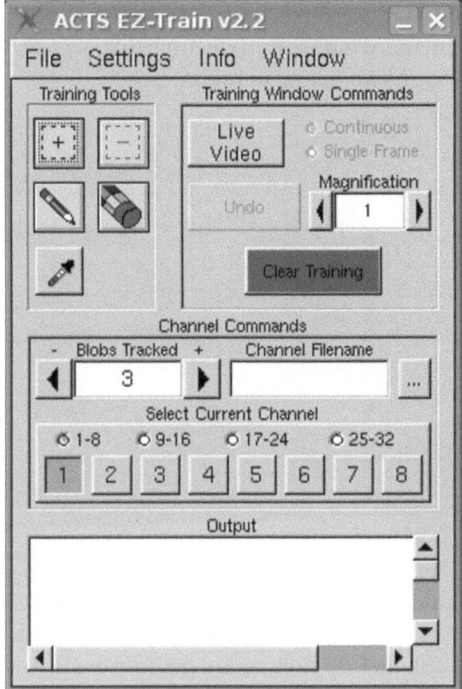

Fig. 3.2 EZ-train control pose window

this file is created save it, for example as "actsconfig" in your robot home directory and then you can start the ACTS server on the robot by typing:

```
/usr/local/acts/bin/acts.bin -t actsconfig &
```

from your home directory. The ACTS server will run, but no channels will be loaded. The -t extension runs ACTS with no graphics, i. e. only the server runs and no attempt is made to load up the EZ-train windows. You can then run savClient on the remote PC by typing:

```
/usr/local/acts/SAV/savClient
```

from the remote PC. A message box will ask which host you want to connect to. Type the name of the robot into the box and you should then see live video images from the robot camera on your remote PC. Use a program such as KSnapshot to capture a number of images of the cans, in different lighting conditions. You should also obtain some images of the background and other colours that the robot can see, so that you can make sure that these colours are not included in the .lut file.

Once you have saved your images you can close ACTS on the robot. On the remote PC load the first picture into the image window by selecting *File → Load Image* from the menu. An example image is shown below in Figure 3.3.

Fig. 3.3 The red cans as seen by the robot

Next select the channel you wish to train by clicking on one of the numbered buttons in the control panel window, see Figure 3.2. Then indicate the maximum number of blobs you want to track on that channel by clicking on the forward arrow in the text box marked "Blobs tracked", or just type the number in; the maximum is 10 for each channel. Colours from the image are added to the channel by selecting pixels or groups of pixels from it. This is done by clicking the "+" button from the "training tools" frame and then clicking and dragging the mouse to create a rectangle that selects the part of the image you want to include. You can repeat this process as many times as possible, adding more and more pixel colours to the channel. If you want to remove pixels from the channel click the "-" button from the "training tools" frame and then select the pixels you want to remove. For example, you might want to do this if you discover that some pixels on the cans also appear in the background.

You can select a number of different image modes; visible mode (Figure 3.4), thresh mode (Figure 3.5) and overlay mode (Figure 3.6). These help to show which pixels are currently recognised by the channel; right clicking the mouse button on the image cycles through them. Visible mode shows the recognised pixels in their actual colours against a black background, thresh mode shows the recognised pixels in white against a black background and overlay mode shows the recognised pixels in blue against the actual background. The blue colour is the default but can be changed to green or red if desired.

A number of images taken in different lighting conditions should be processed to train the channel adequately. If the colour to be tracked does not differ significantly from the background or other colours that the robot may see, it may be quite difficult to prevent the robot detecting false blobs. For example, the background carpet in Figure 3.6 has several shades of pink that may be the same as some of the pink pixels found on the cans. In the image, the cans are not fully shaded blue as many pixels have been removed from the channel to prevent the background carpet and

Fig. 3.4 Visible mode

Fig. 3.5 Thresh mode

Fig. 3.6 Overlay mode

walls of the robot's environment (see Figure 3.7) being tracked as blobs. In general, colours that stand out well from the background should be chosen. It is also better if the objects to be tracked are matte rather than gloss as this prevents reflections from light sources and other colours.

Fig. 3.7 Showing the Pioneer robot, the cans and the background colours

When a channel has been trained it is saved by selecting *File → Save Channel* from the control panel window. A name (with the .lut extension) and a file location can then be specified. Other channels can then be trained and saved as separate .lut files. When all the channels required have been saved a configuration file needs to be created by selecting *File → Save Runtime Config* from the control panel window. All this file does is store the locations of the saved .lut files. Once you have saved all the lookup tables and the configuration file you can close ACTS on the remote PC. The configuration file will have the following format:

```
1 ~/channel1.lut
1 ~/channel2.lut
1 <NoChannel>
1 <NoChannel>
    :
    :
    :
1 <NoChannel>
1 <NoChannel>
1 <NoChannel>
1 <NoChannel>
```

The configuration file shown above has two trained channels 1 and 2; the other 30 channels are not used. The first line specifies the location of channel 1's .lut file "channel1.lut" and the second line specifies the location of channel 2's .lut file "channel2.lut". Here they are stored in the user's home directory. If you need to go back and edit a channel after closing ACTS you can reload the configuration file by typing for example:

```
/usr/local/acts/bin/acts.bin -c actsconfig.
```

The -c parameter instructs ACTS to load the desired configuration file. Note that -t was used on the robot PC as this parameter suppresses loading the graphical training software as well.

Once you have suitable colour lookup tables and a configuration file, copy them to the robot PC. You may need to edit the configuration file as this needs to show where the lookup tables are located in the robot's file system. You can then run ACTS on the robot and load the desired channels by typing for example:

```
/usr/local/acts/bin/acts.bin -t ~/actsconfig
```

where "actsconfig" is the name of the ACTS configuration file saved in your home directory. You can then refer to the channels in your high level code as it is the client programs that decide what to do with the blob tracking information. It is worth mentioning that on start-up the ACTS server consults a file called Acts.pref that is set-up and stored in a hidden directory called .ActivMedia in the user's home directory when they first run ACTS. This sets a number of options and by default it has the following settings:

```
; ACTS preference file. Lines starting with ; are comments


[ACTS]
ImageFilename
ConfigurationFilename
MinimumRunLengthWidth 5
ImageWidth 160
ImageHeight 120
SocketAddr 5001
FrameGrabberType v4l
FrameGrabberDev /dev/video0
FramesPerSec 30
ShowGraphics true
PALCamera false
PXCGrabber false
InvertImage false
FrameGrabberChannel 1
```

The file is a plain text file and can be readily edited. The FrameGrabberChannel variable should be given value 0 for the Pioneers as their cameras use the Composite1 protocol. MinimumRunLengthWidth is the minimum width in pixels for a coloured area to be tracked as a blob. Changing this to 20, for example, would mean that a blob had to consist of 20 or more pixels before it was recognised for tracking. However, the command line arguments supplied when ACTS is run can override the settings of "Acts.pref", see Table 1 of [16] for a full listing of the command line arguments.

3.1.2 Programming ACTS Using ARIA

The ArACTS_1_2 class is used to represent an ACTS object. It is instantiated as described in Section 2.2.2. The class's openPort() method both adds the robot to the device and sets up communication between the robot and the ACTS server. To obtain the number of blobs found on a particular channel use ArACTS_1_2::getNumBlobs(), supplying an integer (the channel number) as the argument. To refer to blobs found by ACTS an instance of an ArACTSBlob class first needs to be declared. The ArACTS_1_2::getBlob() method can then be called. This takes three arguments; the channel number, the blob number and a pointer to the object that will hold its data. You can then obtain its data by calling, for example, the ArACTSBlob::getArea(), ArACTSBlob::getXCG() or ArACTSBlob::getYCG() methods. The first of these returns the area of the blob as an integer value. The last two are used to determine where the blob's centre of gravity is relative to the centre of the camera. The getXCG() method gives the x-coordinate and getYCG() gives the y-coordinate, which are double values. In addition there are the getLeft(), getRight(), getTop() and getBottom() methods which return the positions of the borders of the blob. Before executing a high level client control program that involves ArACTS_1_2 and ArACTSBlob objects the ACTS server must first be running on the robot PC. The program extract below shows how to instantiate an ArACTS_1_2 object and connect it to a robot running ACTS.

```
ArRobot robot;               //Instantiate robot
ArACTS_1_2 acts;             //Instantiate ACTS object
acts.openPort(&robot);       //Add robot and set up communication
                             //with ACTS server running on that robot
```

The user-written method below shows ACTS tracking a colour defined by a particular channel. It returns the value of the largest blob found:

```
/*
*-----------------------------------------------------------
* Blob tracking mode
*-----------------------------------------------------------
*/

int trackBlobs(int channel, double speed, ArACTS_1_2 *thisACTS,
     ArVCC4 *thisPTZ, bool movePTZ)
{

ArACTSBlob blob;                         //Instantiate blob object
ArACTSBlob largestBlob;                  //Instantiate blob object to
int numBlobs;                            //hold largest no. of blobs
int blobArea = 0;                        //Area of blob
bool found = false;                      //Whether blob is found
double xRel;                             //x-co-ord (centre of gravity)
double yRel;                             //y-co-ord (centre of gravity)

numBlobs = thisACTS->getNumBlobs(channel); //Get number of blobs seen

if (numBlobs!=0)                         //If there are blobs
  {
    cout << "Found " << numBlobs << " blobs\n";
    for (int i = 0; i < numBlobs; i++)     //Loop through all blobs
      thisACTS->getBlob(channel, i+1, &blob);
      if (blob.getArea() > blobArea)       //Get area of each blob and
        {                                  //compare with current largest
          found = true;                    //Blob has been found
          blobArea = blob.getArea();       //Set value of largest
          largestBlob = blob;              //Assign largest for tracking
        }
  }else
  {
    cout << "No blobs detected - remaining still\n";
    found = false;
    thisRobot->setVel(0);                  //Set speed to zero
  }

if (found == true )                      //Found blob to track
  {
    /* Determine where the largest blob's center of gravity */
    /* is relative to the center of the camera*/
    xRel = (double)(largestBlob.getXCG() - 160/2.0) / (double)160;
    yRel = (double)(largestBlob.getYCG() - 120/2.0) / (double)120;

    if (movePTZ == true)
      {
        if(!(ArMath::fabs(yRel) < .20))     //Tilt camera toward blob
          {
                                            //Camera moves up or down to
            if (-yRel > 0)                  //centre on blob
              {
                cout << "Tilting camera up toward blob\n";
                thisPTZ->tiltRel(1);
```

```
        }else
        {
          cout << "Tilting camera down toward blob\n";
          thisPTZ->tiltRel(-1);
        }
      }
  }

  // Set the heading for the robot

  if (ArMath::fabs(xRel) < .10)          //If blob central don't adjust
    {
      thisRobot->setDeltaHeading(0);     //XRel should be > 0.1
    }else
    {
      if (ArMath::fabs(-xRel * 10) <= 10) //If blob central
        {                                 //Move in required direction
          thisRobot->setDeltaHeading(-xRel * 10);
        }else if (-xRel > 0)             //If blob is not central
        {
          thisRobot->setDeltaHeading(10);   //Move in required direction
        }else
        {
          thisRobot->setDeltaHeading(-10);
        }
    }
    thisRobot->setVel(speed);                //Set speed for travel to blob
  }
return largestBlob.getArea();               //Return value of largest blob
}
```

3.2 MobileSim

MobileSim simulates MobileRobots platforms and their environments, which is useful for debugging and testing ARIA clients. It is a modification of the Stage simulator (see Chapter 6) created by Richard Vaughan, Andrew Howard and others as part of the Player/Stage project, converting Mapper3Basic .map files (see Section 3.3) to the Stage environment and placing a simulated robot model there. Control is provided via TCP port 8101.

The binary is run from the command line by typing MobileSim. If no additional parameters are specified a dialogue box is opened, see Figure 3.8. This allows you to select your robot type from the *Robot Model* list box (p3dx is the default) and load a map by clicking the *Load Map* button and selecting a saved Mapper3Basic map. Alternatively, the *No Map* button can be clicked. If no map is specified the usable universe (indicated by a grey colour) is limited to 200 metres by 200 metres.

You can also open a map and specify a robot type from the command line by typing:

```
MobileSim -m <map file> -r <robot model>,
```

for example,

```
MobileSim -m mymap.map -r p3dx.
```

If you launch the application in this way no initial dialogue box is displayed.

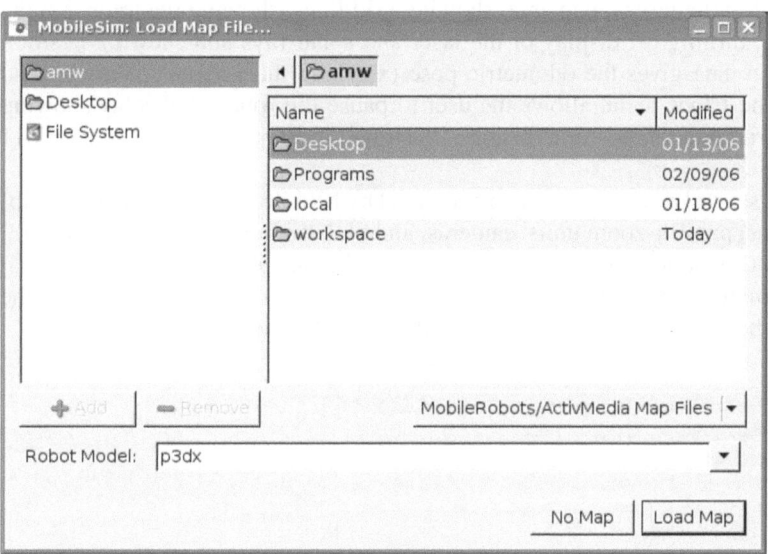

Fig. 3.8 The initial dialogue box for MobileSim

The MobileSim window is opened once the robot type and map have been specified, see Figure 3.9. The map environment and robot are displayed in the centre of the window with the robot at a home position (if this was specified when the map was created) or at the centre. You can pan the window by holding down the right mouse button and dragging and can zoom it with the mouse scroll wheel or by holding down the middle mouse button and dragging towards or away from the centre of the circle that appears. The robot can be moved by dragging it with the left mouse button and can be rotated by dragging with the right mouse button. Both of these actions update the robot's odometry. Grid lines may be added by checking *View →* *Grid* from the menu.

A control program that uses the ArSimpleConnector class to connect to a robot will work on the MobileSim simulator without requiring any modification. This is because the class first tries to connect to MobileSim and only tries to connect to a real robot on a serial connection if MobileSim is not running. A program that uses ArTcpConnection should also work on the simulator with no modification. To run these programs on the simulator you need only run MobileSim and then type the name of the program's binary into the command line, for example `./test`. Figure 3.9 shows the execution of a wandering and obstacle avoidance program that uses the laser and sonar devices. The area shaded blue represents the laser output and the sonar rays are shown in grey coming from the edge of the robot.

The *File* menu allows the user to load a fresh map (*Load File*), reset the robot to its original position on the map (*Reset*) and export frames or sequences of frames (*Export*). The format for frame export and the duration of the export can also be set. The *View* menu allows various display features to be turned off and on. These include shading the laser range area, showing grid lines, showing the trails that the robot makes, turning off display of the laser and sonar rays and showing position data. Position data gives the odometric pose (x, y and theta values), velocity and true pose. The *Clock* menu allows the user to pause the robot. A display showing the robot's trail and the position data are illustrated in Figure 3.10 and Figure 3.11 respectively.

Note that several devices cannot be simulated by MobileSim. These include grippers, 5D arms, pan-tilt-zoom units, cameras, and blob finding devices, see Table 1.2 for a full list. MobileRobots does not have any immediate plans to update MobileSim to include these devices, but it is likely that a version that includes the gripper will be released before any version that includes the blob finder.

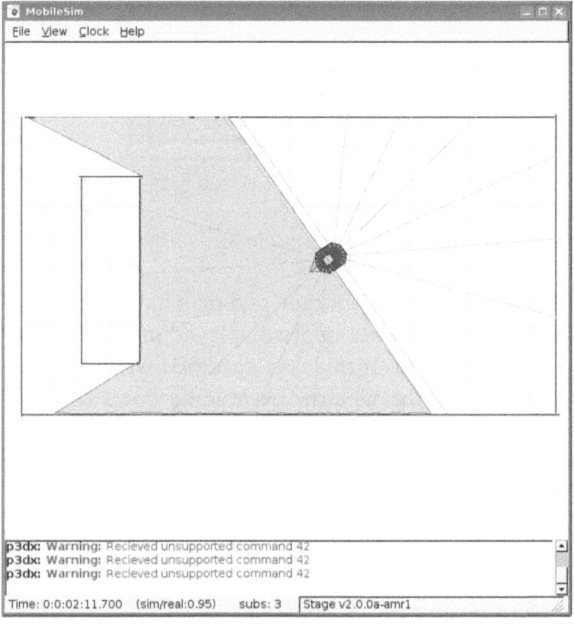

Fig. 3.9 The MobileSim GUI window

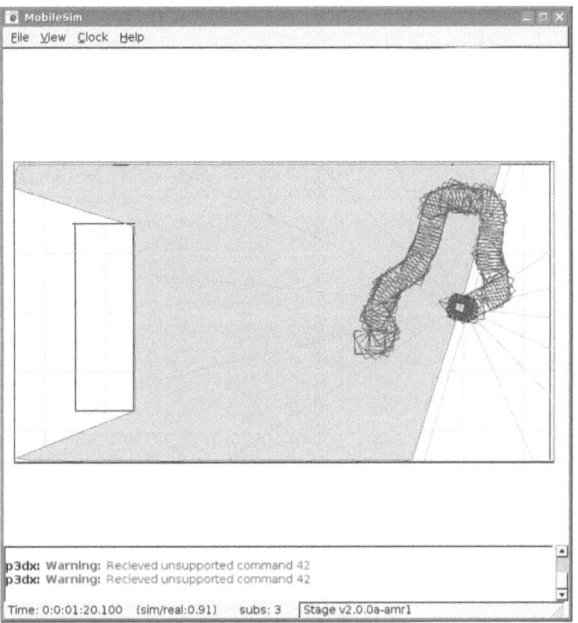

Fig. 3.10 The simulated Pioneer's trail

3.3 Mapper3Basic

Mapper3Basic can be used to create and edit maps for MobileSim (see Section 3.2) so that walls and other obstacles can be simulated. This can be done by drawing map lines, goals, forbidden lines and areas, home points and areas and dock points.

The binary is run from the command line by typing Mapper3Basic, which opens a graphical window shown in Figure 3.12. To start a new map select *File → New* from the menu and a blank sheet will be loaded. To open an existing map select the *Open* icon or *File → Open* from the menu. If you require grid lines you can select *View → Grid Lines* from the menu.

Lines, goals and other map objects are placed on the sheet by selecting the appropriate button from the second row and then clicking and dragging the mouse to draw the object. The example above shows four lines drawn to form a rectangle and another four drawn to form an inner rectangle (unshaded). If placed outside the inner rectangle but inside the outer rectangle the robot would be able to move within the outer but would not be able to enter the inner rectangle. However, this is not a truly forbidden area as the robot could be placed within the inner rectangle and would still be free to move around. Forbidden areas are created by selecting the *Forbidden Area* icon and clicking and dragging the mouse over the area that the robot must not enter. These areas are shown shaded orange. In addition, forbidden lines can also be created using the *Forbidden Line* icon. These could be used to prevent the robot from

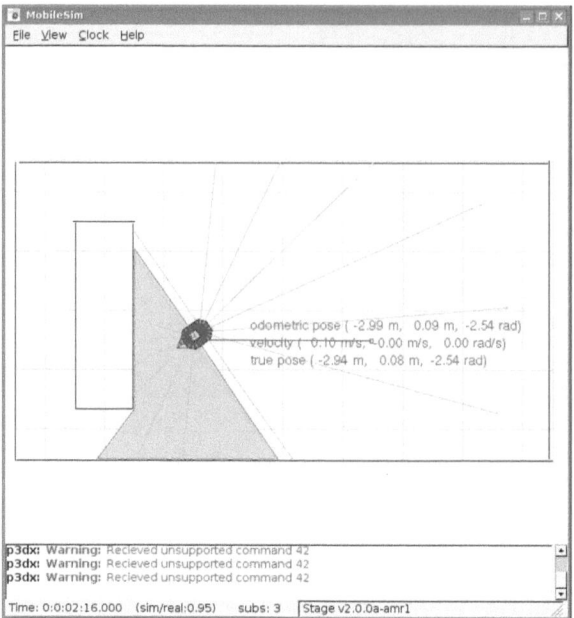

Fig. 3.11 The simulated Pioneer's position data

getting too close to hazards that cannot be detected with range sensors, for example staircases and holes. If you require your robot to avoid forbidden areas you will also need to create an instance of a virtual ranged device ArForbiddenRangeDevice in your ARIA program and add it to the robot. This is used to measure the distances from forbidden areas.

If you require your robot to begin in a particular location on the map then select the *Home Point* icon and click on the point where the robot must begin. Maps are saved as bitmap images in the form of .map files by selecting the *Save* icon or choosing *Save* or *Save As* from the file menu. Once saved the maps can be loaded into MobileSim.

Goals, home areas and dock points can also be created. However, these features are for use when creating maps for MobileEyes, MobileRobots' GUI navigation system for remote robot control and monitoring. MobileEyes can connect to ARIA, ArNetworking and ARNL (ARIA's Navigation Library) servers over a wireless network to display the map of the robot's environment. It provides controls to send the robot to goal points or any other point on the map, and also allows the robot to be driven directly with the keyboard or joystick. However, further details about MobileEyes and the navigation library are not included in this guide as details about MobileEyes are available with the online documentation that comes with the software.

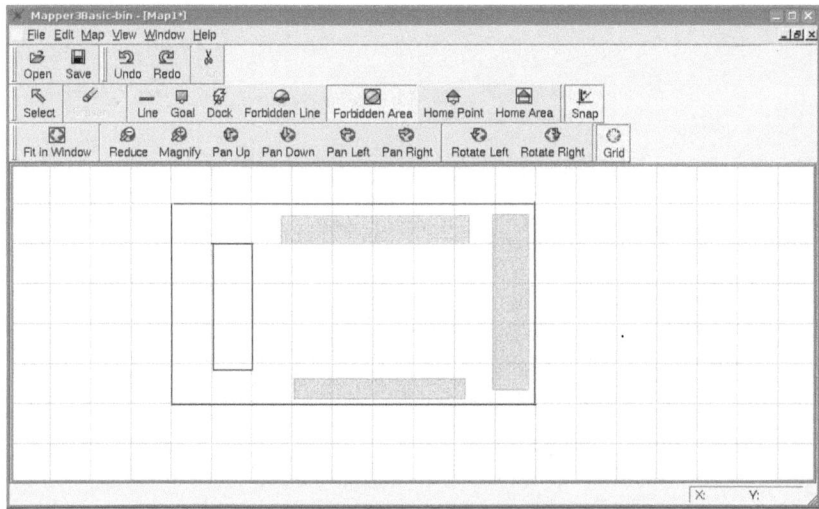

Fig. 3.12 The interface for Mapper3Basic

The next chapter examines the use of subclasses within ARIA and covers each of ArAction, ArActionGroup, and ArMode subclasses.

Chapter 4
Using ARIA Subclasses

4.1 Creating and Using ArAction Subclasses

Another way of controlling a robot with ARIA is to create action subclasses that inherit from the base ArAction class. When instances of these classes are added to an ArRobot object the robot's resulting behaviour is determined through an action resolver. This invokes each ArAction object (via its fire() method), and the actions request what kind of motion they want by returning a pointer to an ArActionDesired object. The action resolver determines what the resulting combination of those requested motions should be, then commands the robot accordingly. The idea behind this is to have several behaviours acting simultaneously, which combine to drive the robot.

When using the ArAction class direct commands can still be used (for example ArRobot::setVel()), but if you mix direct motion commands with ArAction objects you must fix ArRobot's state by calling ArRobot::clearDirectMotion() before actions will work again.

The program below shows how to create an action that inherits from the ArAction class. This is an adaptation of the actsSimple.cpp program that appears in /usr/local/Aria/examples. Note that it is similar to the blob finding method shown in Section 3.1.2. The difference is that this is a class inheriting from ArAction, whereas the program in Section 3.1.2 was just a method.

```
#include "Aria.h"
#include <iostream>

/* This class moves a robot toward the largest blob seen */

class Blobfind : public ArAction                                    1
{
public:

  enum State                              //State of action
    {
      NO_TARGET,                          //No target in view
```

```
    TARGET,                                      //Target in view
  };

  Blobfind(ArACTS_1_2 *acts, ArVCC4 *camera); //Constructor        2
  ~Blobfind(void);                             //Destructor
  ArActionDesired *fire(ArActionDesired currentDesired);           3
  State getState(void)  return myState;     //Return state of action

protected:                                                         4

  ArActionDesired myDesired;                                       5
  ArACTS_1_2 *myActs;
  ArVCC4 *myCamera;
  State myState;
  int myChannel;
};

                                             // Constructor
Blobfind::Blobfind(ArACTS_1_2 *acts, ArVCC4 *camera) :            6
  ArAction("Blobfind", "Moves towards the largest blob.")

{
  myActs = acts;
  myCamera = camera;
  myChannel = 1;
  myState = NO_TARGET;
}

Blobfind::~Blobfind(void)                      //Destructor

// The fire method
ArActionDesired *Blobfind::fire(ArActionDesired currentDesired)   7
{
  ArACTSBlob blob;
  ArACTSBlob largestBlob;

  bool flag = false;
  int numberOfBlobs;
  int blobArea = 10;
  double xRel, yRel;

  myDesired.reset();                           //Reset desired action   8

  numberOfBlobs = myActs->getNumBlobs(myChannel);

  if(numberOfBlobs != 0)                       //If there are blobs
    {
      myState = TARGET;
      for(int i = 0; i < numberOfBlobs; i++)
        {
          myActs->getBlob(myChannel, i + 1, &blob);
          if(blob.getArea() > blobArea)
            {
              flag = true;
```

```
                blobArea = blob.getArea();
                largestBlob = blob;
              }
        }
    }else
    {
      myState = NO_TARGET;
    }

  if(flag == true)
    {
      //Determine where the largest blob's center of gravity
      //is relative to the center of the camera
      xRel = (double)(largestBlob.getXCG() - 160.0/2.0) / 160.0;
      yRel = (double)(largestBlob.getYCG() - 120.0/2.0) / 120.0;

      if(!(ArMath::fabs(yRel) < .20))           //Tilt camera toward blob
        {
          if (-yRel > 0)
            myCamera->tiltRel(1);
          else
            myCamera->tiltRel(-1);
        }

      if (ArMath::fabs(xRel) < .10)             //Set heading and velocity
        {
          myDesired.setDeltaHeading(0);                                 9
        }
      else
        {
          if (ArMath::fabs(-xRel * 10) <= 10)
            myDesired.setDeltaHeading(-xRel * 10);                      10
          else if (-xRel > 0)
            myDesired.setDeltaHeading(10);                             11
          else
            myDesired.setDeltaHeading(-10);                            12
        }

      myDesired.setVel(200);                                          13
      return &myDesired;                                              14
    }
  else
    {
      myDesired.setVel(0);                                            15
      myDesired.setDeltaHeading(0);                                   16
      return &myDesired;                                              17
    }
}
```

The important lines in the program are numbered. Line 1 declares the class as a subclass of ArAction. Line 2 declares the constructor, which takes pointers to an ACTS device and a camera as its arguments. Line 3 declares the fire() method, which is the important one to override for subclasses of ArArction. It must return a

pointer to an ArActionDesired object to indicate what the action wants to do and can
be NULL if the action does not want to change what the robot is currently doing. It
must also have ArActionDesired currentDesired as its parameter. This enables the
action to determine what the resolver currently wants to do as currentDesired refers
to the resolver's current desired action. It is used solely for the purpose of giving
information to the action.

Line 4 begins declaration of the protected attributes of the class; only subclasses
have access to these. Line 5 declares the ArActionDesired object called "myDesired". Line 6 begins the constructor method and the right hand side part, for example : ArAction("Blobfind", "Moves towards the largest blob") must be included.
Line 7 begins the fire() method. This method sets the action request by returning
the pointer to "myDesired". Line 8 resets "myDesired" and lines 9 to 17 set "my-
Desired" under different conditions. Note that "myDesired" is used with ArRobot
direct motion commands like setDeltaHeading().

A main method that uses the above action is given below. This assumes that the
above class was saved as "BlobFind.cpp".

```
#include "Aria.h"
#include "BlobFind.cpp"                                                  1
#include <stdio.h>
#include <iostream>

using namespace std;

int main(int argc, char** argv)
{
  ArRobot robot;                        //Instantiate robot
  ArSonarDevice sonar;                  //Instantiate sonar
  ArVCC4 vcc4 (&robot);                 //Instantiate camera
  ArACTS_1_2 acts;                      //Instantiate acts device
  ArSimpleConnector simpleConnector(&argc, argv);

  if (!simpleConnector.parseArgs() || argc > 1)
    {
      simpleConnector.logOptions();
      exit(1);
    }

  /* Instantiate actions */
  ArActionLimiterForwards limiter("speed limiter near", 300,600,250);    2
  ArActionLimiterForwards limiterFar("speed limiter far", 300,1100,400); 3
  ArActionLimiterBackwards backwardsLimiter;                             4
  ArActionConstantVelocity stop("stop", 0);                             5
  ArActionConstantVelocity backup("backup", -200);                      6
  Blobfind blobFind(&acts, &vcc4);     //Blob finding action            7

  Aria::init();
  robot.addRangeDevice(&sonar);         //Add sonar to robot

  /* Connect to the robot */
  if (!simpleConnector.connectRobot(&robot))
```

```
  {
    cout << "Could not connect to robot... exiting\n";
    Aria::shutdown();
    return 1;
  }

acts.openPort(&robot);               //Connect to acts
vcc4.init();                         //Initialise camera
ArUtil::sleep(1000);                 //Wait a second.....
robot.setAbsoluteMaxTransVel(400);
robot.comInt(ArCommands::ENABLE, 1); //Enable motors
ArUtil::sleep(200);

/* Add actions to robot */
robot.addAction(&limiter, 100);                             8
robot.addAction(&limiterFar, 99);                           9
robot.addAction(&backwardsLimiter, 98);                    10
robot.addAction(&blobFind, 77);                            11
robot.addAction(&backup, 50);                              12
robot.addAction(&stop, 30);                                13
robot.run(true);                     //Run the program     14

Aria::shutdown();
return 0;
}
```

Here, line 1 includes the file containing the ArAction subclass "Blobfind". Lines 2 to 7 declare instances of the action classes that the robot will use; line 7 is the "Blobfind" action created earlier, the others are all standard ArAction subclasses that form part of the ARIA library. ArActionLimiterForwards and ArActionLimiter-Backwards limit the forwards and backwards motion of the robot respectively based on range sensor readings, and ArActionConstantVelocity simply sets the robot at a constant velocity. Lines 8 to 14 add the actions to the robot using ArRobot's ad-dAction() method. This method takes a pointer to an ArAction object and an integer value representing the action's priority as its arguments. The priority values are used by the action resolver to determine the final desired action of the robot. Line 14 runs the program.

4.2 Creating and Using ArActionGroup Subclasses

ArActionGroup subclasses are used to wrap a group of ArAction subclasses together to form an action group. This is useful if you have a number of actions that implement a behaviour collectively but you want to be able to activate the behaviour with one call to the group's activate() method. The program below shows how to group actions using a subclass of the ArActionGroup base class. It does the same job as the previous example, i.e. carries out blob tracking at the same time as limiting the forward and backward robot motions. The difference is that here all the actions are

added to the robot and their priorities are set using the addAction() method of the
ArActionGroup class. The action "blobFind" created in the previous section is im-
plemented by including the source file (line 1), declaring a pointer to an instance of
the class (lines 2 and 3) and adding the action to the group (line 4).

```cpp
#include "Aria.h"
#include "BlobFind.cpp"                                                        1

/* Implements blob finding in an action group */

class BlobfindGroup : public ArActionGroup
{
public:
  BlobfindGroup(ArRobot *robot, ArACTS_1_2 *acts, ArVCC4 *camera);
  ~BlobfindGroup(void);

  /* Declaring the actions that will be added */

protected:
  Blobfind* blobFind;                                                         2
  ArActionLimiterForwards* limiter;
  ArActionLimiterForwards* limiterFar;
  ArActionLimiterBackwards* backwardsLimiter;
  ArActionConstantVelocity* stop;
  ArActionConstantVelocity* backup;
};

/* Constructor */

BlobfindGroup::BlobfindGroup(ArRobot *robot, ArACTS_1_2 *acts,
          ArVCC4 *camera):ArActionGroup(robot)
{

  /* Instantiate and add the actions to the group */

  blobFind = new Blobfind(acts, camera);                                      3
  limiter = new ArActionLimiterForwards("speed limiter near", 300, 600,
          250);
  limiterFar = new ArActionLimiterForwards("speed limiter far", 300, 1100,
          400);
  backwardsLimiter = new ArActionLimiterBackwards;
  stop = new ArActionConstantVelocity("stop", 0);
  backup = new ArActionConstantVelocity("backup", -200);

  addAction(blobFind, 77);                                                    4
  addAction(limiter, 100);
  addAction(limiterFar, 99);
  addAction(backwardsLimiter, 98);
  addAction(backup, 50);
  addAction(stop, 30);
}

/* Destructor */

BlobfindGroup::~BlobfindGroup(void) {}
```

The main method (below) remains largely unchanged, except that there is now no need to declare and add the actions as they are already declared in the group class. All you need to do is declare the group (line 1) and then activate it exclusively (line 2).

```cpp
#include "Aria.h"
#include "BlobFindGroup.cpp"
#include <stdio.h>
#include <iostream>

using namespace std;

int main(int argc, char** argv)
{

  ArRobot robot;                               //Instantiate robot
  ArSonarDevice sonar;                         //Instantiate sonar
  ArVCC4 vcc4 (&robot);                        //Instantiate camera
  ArACTS_1_2 acts;                             //Instantiate acts
  ArSimpleConnector simpleConnector(&argc, argv);

  if (!simpleConnector.parseArgs() || argc > 1)
    {
      simpleConnector.logOptions();
      exit(1);
    }

  Aria::init();
  robot.addRangeDevice(&sonar);                //Add sonar to robot

  /* Declare the Blobfinder action group */

  BlobfindGroup blobFindGroup(&robot, &acts, &vcc4);               1

  /* Connect to the robot */
  if (!simpleConnector.connectRobot(&robot))
    {
      cout << "Could not connect to robot... exiting\n";
      Aria::shutdown();
      return 1;
    }

  acts.openPort(&robot);                       //Connect to acts
  vcc4.init();                                 //Initialise camera
  ArUtil::sleep(1000);                         //Wait a second...
  robot.setAbsoluteMaxTransVel(400);           //Set max speed
  robot.comInt(ArCommands::ENABLE, 1);         //Enable motors
  ArUtil::sleep(200);

  /* Implement the actions in the group */
  blobFindGroup.activateExclusive();                              2
```

```
robot.run(true);                        //Run the program

Aria::shutdown();
return 0;
}
```

4.3 Creating and Using ArMode Subclasses

Once an ArActionGroup subclass has been created an ArMode subclass that implements the group action using a single keyboard character can be written. The program below shows how to implement this. Line 1 includes the "BlobfindGroup" class and line 2 declares an instance of the class. The constructor just deactivates the group behaviour initially, line 3. The activate() and deactivate() methods of the base class (starting on lines 4 and 5 respectively) must be overridden in the subclass. These methods should just activate and deactivate the behaviour.

```
#include "Aria.h"
#include "BlobFindGroup.cpp"                                        1

class BlobfindMode : public ArMode
{
public:
  BlobfindMode(ArRobot *robot, const char *name, char key, char key2,
           ArACTS_1_2 *acts, ArVCC4 *camera);
  ~BlobfindMode(void);
  void activate();                        //Activate mode
  void deactivate();                      //Deactivate mode

/* Declaring the group associated with this mode */

protected:
  BlobfindGroup myGroup;                                            2

};

/* Constructor */

BlobfindMode::BlobfindMode(ArRobot *robot, const char *name, char key,
      char key2, ArACTS_1_2 *acts, ArVCC4 *camera) : ArMode(robot,
      name, key, key2),
myGroup(robot, acts, camera)
  {
    myGroup.deactivate();                 //Deactivate group        3
  }                                       //(only run when key pressed)

/* Destructor */

BlobfindMode::~BlobfindMode(void) {}
```

```
void BlobfindMode::activate()                                       4
{
if (!baseActivate())
  {
    return;
  }
myGroup.activateExclusive();
}


/* Deactivation method */

void BlobfindMode::deactivate()                                     5
{
if (!baseDeactivate())
  {
    return;
  }
myGroup.deactivate();
}
```

A main method that implements the blob finding mode is shown below. It also has a default wander mode. The user can switch into blob finding mode by pressing the "b" or "B" key.

```
#include "Aria.h"
#include "BlobFindMode.cpp"                                         1
#include <stdio.h>
#include <iostream>

using namespace std;

int main(int argc, char** argv)
{

  ArRobot robot;                                //Instantiate robot
  ArSonarDevice sonar;                          //Instantiate sonar
  ArVCC4 vcc4 (&robot);                         //Instantiate camera
  ArACTS_1_2 acts;                              //Instantiate acts
  ArSimpleConnector simpleConnector(&argc, argv);
  ArKeyHandler keyHandler;                      //Instantiate key handler   2

  if (!simpleConnector.parseArgs() || argc > 1)
    {
      simpleConnector.logOptions();
      exit(1);
    }

  Aria::init();
  robot.addRangeDevice(&sonar);                 //Add sonar
  Aria::setKeyHandler(&keyHandler);                                         3
  robot.attachKeyHandler(&keyHandler);          //Add key handler           4

  /* Connect to the robot */
  if (!simpleConnector.connectRobot(&robot))
```

```
  {
    cout << "Could not connect to robot... exiting\n";
    Aria::shutdown();
    return 1;
  }

acts.openPort(&robot);                          //Connect to acts
vcc4.init();                                    //Initialise camera
ArUtil::sleep(1000);                            //Wait a second...
robot.setAbsoluteMaxTransVel(400);              //Set max speed

ArUtil::sleep(200);
robot.runAsync(true);

/* Set up robot */

robot.lock();
robot.comInt(ArCommands::ENABLE, 1);            //Enable motors

/* Implement blobfinding mode or wander mode using single keys */

BlobfindMode blobFind(&robot, "blobfind-mode", 'b','B', &acts, &vcc4); 5
ArModeWander wander(&robot, "wander-mode", 'w', 'W');                  6

wander.activate();                              //Set default behaviour   7

robot.unlock();
robot.waitForRunExit();                         //Run the program          8
Aria::shutdown();
return 0;
}
```

Line 1 includes the file containing the "BlobfindMode" class. Line 2 instantiates an ArKeyHandler object for processing key events and lines 3 and 4 register it with the main ARIA library and attach it to the ArRobot object respectively, which should be done before connecting to the robot. Line 5 declares a "BlobfindMode" object. Note that the parameters supply the name of the mode and the keys that will be used to switch to it. Line 6 does the same for the "ArModeWander" mode that is part of the ARIA library. Line 7 sets the default behaviour to wandering, i.e., the robot will begin with this behaviour when the program starts. Line 8 runs the program.

The next chapter deals with programming Pioneer robots using Player software, and as with the ARIA section, each device is treated separately.

Chapter 5
Programming with Player

5.1 Player Configuration Files

In order to run Player you need to specify a configuration file, which is a simple text file that tells the robot which drivers to use for which device. If you placed your original Player tarball in /usr/src then the source code for all of the drivers will have been placed in /usr/src/player-2.0.5/server/drivers when you unzipped it, so you can browse through these files to see which drivers are potentially available. All the drivers that were built on your system at the time of installation will have been compiled and linked into "libplayerdrivers", but not all drivers are built by default, see Section 1.5.4.

In the configuration file the drivers used are specified by name and then a list of the interfaces they provide is given. The port they must communicate with is also listed in some cases and some drivers allow the specification of other attributes. An example configuration file for the Pioneers is presented below:

```
# Pioneer P3-DX-SH

driver
(
    name "p2os"
    provides ["odometry:::position2d:0"
    "sonar:0"
    "gripper:0"
    "power:0"
    "bumper:0"]
    port "/dev/ttyS0"
)

driver
(
```

```
    name "canonvcc4"
    provides ["ptz:0"]
    port "/dev/ttyS1"
)

driver
(
    name "sicklms200"
    provides ["laser:0"]
    port "/dev/ttyS2"
    rate 38400
    delay 32
    resolution 50
    range_res 1
    alwayson 1
)

driver
(
    name "camerav4l"
    provides ["camera:0"]
    port "/dev/video0"
    source 0
    size [320 240]
    norm "ntsc"
    mode "RGB888"
)

driver
(
    name "cmvision"
    requires ["camera:0"]
    provides ["blobfinder:0"]
    colorfile ["colors2.txt"]
)
```

The first driver listed is the p2os. This driver offers access to the micro-controller that runs the special embedded operating system on Pioneer robots. It is connected to the serial port /dev/ttyS0 of the robot and mediates control of several interfaces explained in Table 5.1.

The next driver listed is the canonvcc4, which controls a Canon VC-C4 pan-tilt-zoom unit. The configuration file tells Player that the driver provides the pan-tilt-zoom function and that the unit is connected to serial port /dev/ttyS1. Note that when using Player all pan-tilt-zoom units must be connected to a serial port otherwise they will not function, see Section 5.5.7 for further details.

Table 5.1 P2OS interfaces

Interface	Function
odometry:::position2d	Returns odometry data and accepts velocity commands
sonar	Returns data from sonar arrays (if equipped)
gripper	Controls gripper (if equipped)
power	Returns the current battery voltage (12 V when fully charged)
bumper	Returns data from bumper array (if equipped)

The next driver is the sicklms200, the driver for the SICK LMS200 laser that comes with some Pioneer robots. The configuration file tells Player that the driver provides an interface for the laser and that it is connected to port /dev/ttyS2. In addition, there are a number of other attributes that can be set including the baud rate (the default is 38400), the delay in seconds before initialisation (the default is 0), the resolution (50 = 361 readings, 100 = 181 readings, the default is 50) and the range resolution. The default range resolution is 1, which means 1 mm precision, 8.192 m max range. This can also be set to 10 (10 mm precision, 81.92 m max range) or 100 (100 mm precision, 819.2 m max range). The alwayson attribute can also be used to specify whether to keep the laser on all the time that Player is running. If the parameter 1 is used, the laser remains on. The default value is 0, which means that the laser is disconnected after any client programs that use it have finished executing. In terms of saving battery power it is better not to keep the laser running all the time.

The camerav4l driver captures images from V4l-compatible cameras. The port specified is the device for reading the video data, i.e., the framegrabber /dev/video0 in this case. Other attributes include source, for specifying which capture card input source should be used, norm for specifying NTSC or PAL capture standards and size for indicating the required image size. Mode tells Player the desired capture mode. The options are:

- RGB888 (24-bit RGB) The default
- RGB565 (16-bit RGB)
- RGB32 (32-bit RGB; producing 24-bit colour images)
- YUV420P (planar YUV data converted to 24-bit colour images)
- YUV420P_GREY (planar YUV data; producing 8-bit monochrome)

The last driver listed is the cmvision driver, which provides a virtual blob finder device. The configuration file shows that the device requires the camera and will not function without it. This is expressed by using the requires key word. A "color" or configuration file must also be specified for indicating the colours to be tracked. Section 5.5.8 provides more details on the cmvision blob finder.

Once you have set up a configuration file for your robot and saved it (for example as "pioneer.cfg" in your home directory on the robot) you can run the Player server by opening a secure shell on the robot and typing:

```
player pioneer.cfg &
```

on the command line. This assumes you are in your home directory and that your system knows where the Player binary is. If not then you will need to specify the full path names of Player's binary and the configuration file. When the Player server runs you will see the following output on the command line:

```
* Part of the Player/Stage/Gazebo Project [http://playerstage.sourceforge.net].
* Copyright (C) 2000 - 2005 Brian Gerkey, Richard Vaughan, Andrew Howard,
* Nate Koenig, and contributors. Released under the GNU General Public License.
* Player comes with ABSOLUTELY NO WARRANTY. This is free software, and you
* are welcome to redistribute it under certain conditions; see COPYING
* for details.

Listening on ports: 6665
```

You can also obtain more debugging information by typing:

```
player -d 9 pioneer.cfg &.
```

Note that if there are errors in your configuration file the server will report the error and will not run. Once you have the Player server running on your robot you can run high level client control programs from the robot itself or from a remote PC. You can also run a tool called PlayerViewer remotely to get visual images of the output from the robot's sensors, see Section 5.2 below. To kill the Player server at any time type either `killall player` or `kill -9` followed by the process number.

5.2 Using PlayerViewer

PlayerViewer is a GUI client program that enables a user to view sensor data from a player server. It also provides some teleoperation capabilities. PlayerViewer is installed at the same time as Player, by default in `/usr/local/bin`. To run it you must first run the Player server on a real robot or run Player with Stage, see Chapter 6. You need X-Windows on your system to view the output from PlayerViewer, so when running Player on a real robot it is normally run from a remote PC. If you are running Player with Stage then typing:

```
/usr/local/bin/playerv
```

from a shell console or just `playerv` (if your system knows where the binary is located) will bring up the PlayerViewer GUI interface. If you are running Player on a real robot then from the remote PC shell type:

```
playerv -h <robotname>
```

or

```
playerv -p <portnumber>
```

where <robotname> is the host name of your robot and <portnumber> is the port number that Player is listening on. Figure 5.3, Figure 5.4 and Figure 5.5 below show the GUI interface for PlayerViewer. To pan the window click and drag with the left mouse button. To zoom click and drag with the right mouse button.

The GUI has three root menus, *File*, *View* and *Devices*. The user is able to see the output from the robot's devices by subscribing to them from the *Devices* menu. This menu lists the available devices based on information supplied by the configuration file. To subscribe to a device select *Subscribe* from the sub-menu for that device. Most devices only have *Subscribe* in their sub-menu, but *position2d:(p2os)* also has options to command the robot's position, i.e., move the robot manually. The *File* menu is used to save images either as stills in jpeg or ppm formats, or as movies at one and two times normal speed. It also provides an *Exit* function to close the application. The *View* menu provides options for rotating the image and for turning grid lines off and on.

Figure 5.3 is the PlayerViewer output for a real Pioneer robot in a small pen with two gate posts, see Figure 5.1. The user was subscribed to the position, laser and sonar devices and the *Command* and *Enable* options were selected from the *position2d:(p2os)* sub-menu. The solid red line in Figure 5.3 shows the robot position, the purple area shows the laser range and the brown triangles represent the sonar output. Note how the laser sensor is much more accurate than the sonars in the real world. The dotted red line illustrates use of the command feature. To command the robot's position the target crosshairs inside the robot are dragged to the desired location. *Enable* must be selected from the *position2d:(p2os)* sub-menu for this to work.

Figure 5.4 and Figure 5.5 show PlayerViewer images based on a virtual Stage Pioneer robot in the world shown in Figure 5.2. In Figure 5.4 the user was subscribed to the position, pan-tilt-zoom, gripper, blob finder and sonar devices. Here, the pan-tilt-zoom field of view is shown by the green lines. To pan and zoom the camera unit click and drag the green circle. Clicking and dragging the blue circle will also tilt a real camera. The blobs seen are shown in the small window underneath and to the right of the robot. In Figure 5.5 the user was subscribed to the position, laser, pan-tilt-zoom and gripper devices.

Fig. 5.1 The real environment upon which Figure 5.3 is based

5.3 Programming with the Player C++ Client Library

The remainder of this chapter covers use of the Player C++ library, which provides high level client control programs for real robots and virtual ones created through Player's 2D simulator Stage.

5.3.1 Compiling Programs

Player programs are compiled under Linux by using g++ on the command line and programs must be linked to the Player C++ library, so you may need to add the path /usr/local/lib to the file /etc/ld.so.conf to let your dynamic linker know where the "libplayerc++.so.2" file is. After adding it you will also need to run ldconfig. You may also need to add the same directory to the LD_LIBRARY_PATH environment variable. As an example, suppose you have a control program called "test.cpp" and you wish to create a binary called "test". From the directory where "test.cpp" is located, you would type the following:

```
g++ -o test`pkg-config --cflags playerc++` test.cpp
     `pkg-config --libs player.
```

Alternatively you could write a shell script, such as the one shown below to save

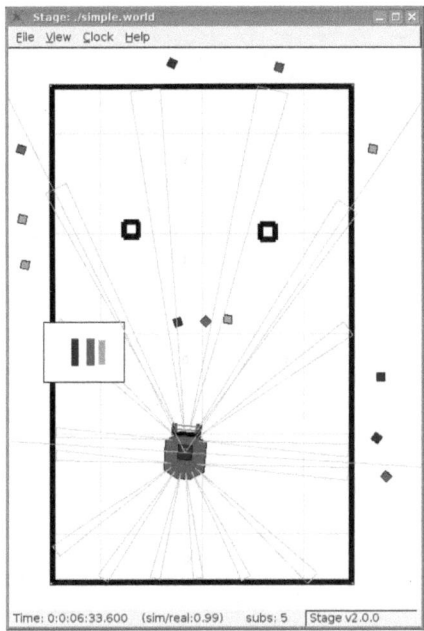

Fig. 5.2 Stage world upon which Figure 5.4 and Figure 5.5 are based

typing this every time you want to compile:

```
#!/bin/sh

# Short script to compile a Player 2.0.5
# playerc++ client
# Requires 2 arguments, (1) name of binary
# and (2) name of program to compile

if [ $# != 2 ]; then
   echo Require 2 arguments
   exit 1
fi

g++ -Wall -o $1 `pkg-config --cflags playerc++` $2
      `pkg-config --libs playerc++`
```

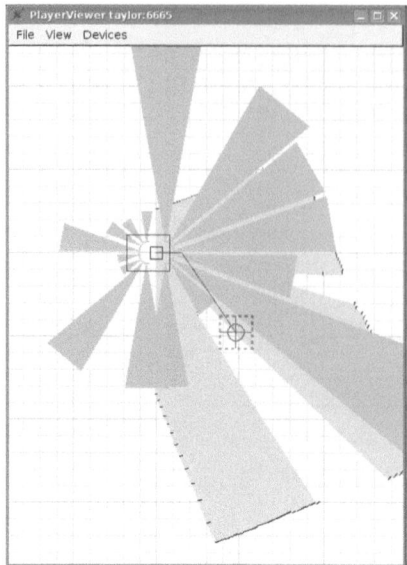

Fig. 5.3 PlayerViewer output for a real Pioneer robot

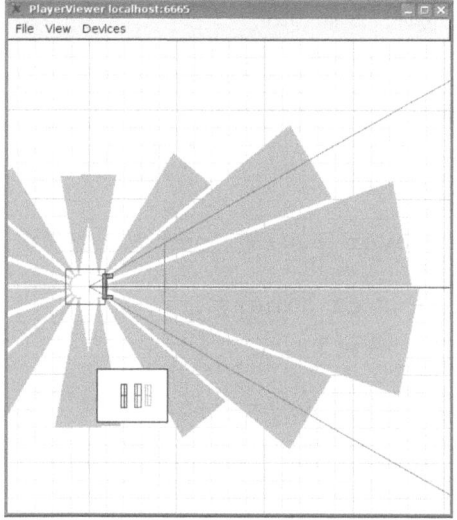

Fig. 5.4 PlayerViewer output for a simulated robot with sonar sensors

5.3.2 Connecting to a Robot

The Player C++ client library uses classes as proxies for local services. There are
two kinds, the single server proxy PlayerClient and numerous proxies for the devices

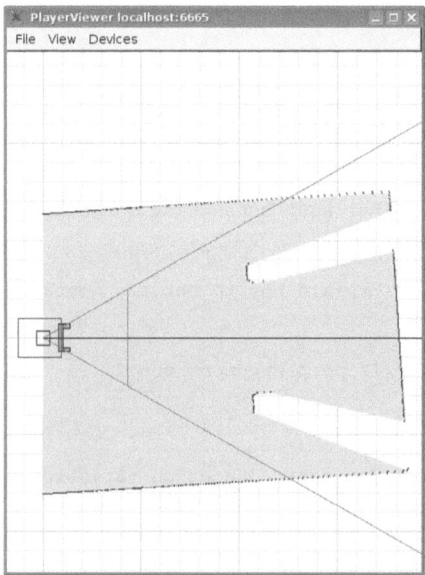

Fig. 5.5 PlayerViewer output for a simulated robot with laser sensor

used, for example the Position2dProxy for obtaining position and setting speed. Connection to a Player server is achieved by creating an instance of the PlayerClient proxy and specifying host and (or) port arguments. The program below shows how to pass and process host and port arguments and create the server proxy. The important lines are numbered.

```
#include <libplayerc++/playerc++.h>      //c++ client library      1
#include <stdio.h>
#include <stdlib.h>
#include <iostream>

using namespace std;
using namespace PlayerCc;                                          2

#define USAGE \                                                    3

  "USAGE: test-program [-h <host>] [-p <port>] \n"
  " -h <host>: connect to Player on this host\n"
  " -p <port>: connect to Player on this TCP port\n"

char host[256] = "localhost";             //Default host name      4
int port = 6665;                          //Default port number    5

int main(int argc, char **argv)
{
/* Set the host and port arguments*/
int i = 1;
```

```
while (i < argc)
  {
    if(!strcmp(argv[i],"-h"))              //If host argument specified
      {
        if(++i < argc)
          {
            strcpy(host,argv[i]);          //Set host connection variable 6
          }else
          {
            puts(USAGE);                   //Explain how to set arguments
            exit(1);
          }
      }else if(!strcmp(argv[i],"-p"))      //If port argument specified
      {
        if(++i < argc)
          {
            port = atoi(argv[i]);          //Set port connection variable 7
          }else
          {
            puts(USAGE);                   //Explain how to set arguments
            exit(1);
          }
      }
    i++;
  }

PlayerClient robot(host, port);           //Create PlayerClient          8
robot.SetDataMode(PLAYER_DATAMODE_PULL); //Set data to PULL mode         9
robot.SetReplaceRule(-1, -1, PLAYER_MSGTYPE_DATA, -1, 1);                10

}
```

Line 1 includes the Player C++ client library and line 2 indicates that the pre-defined PlayerCc namespace is being used. These lines must be included for all Player C++ client programs. Line 3 defines the output that will be displayed on screen if the host and port arguments are not specified in the right way. Lines 4 and 5 set the default values for host and port respectively if no arguments are passed. In the first part of the program the arguments that were passed at run time are examined and, if necessary, they are passed to the host and port variables (lines 6 and 7). This is important since no arguments need to be passed if a Stage simulation is being run. However, if the client is run on a real robot but remotely from a networked PC then a host name or port number needs to be given. Line 8 creates an instance of a PlayerClient object called "robot". Lines 9 and 10 set the data mode for message passing. In PULL mode the server only sends data after the client requests it. Player queues up messages and sends the current contents of the queue upon the client's request. The default mode is PUSH in which Player pushes messages out to the client as fast as it can.

If the compiled client binary of the above program was called "test" you would type either:

```
./test -h <robotname>
```

or

```
./test -p <portnumber>
```

to run the program, where <robotname> and <portnumber> are the host and port numbers respectively. If you are running Stage then just typing ./test should run the client in the Stage window, see Chapter 6.

5.4 Instantiating and Adding Devices

All devices are registered by creating instances of the appropriate proxies and initialising them through the established PlayerClient object. The program extract below shows how to instantiate laser and sonar devices and associate them with a Player-Client object called "robot". Note that connection is achieved here by simply using the argument "localhost".

```
PlayerClient robot(localhost);    //Create instance of PlayerClient   1
SonarProxy sp(&robot, 0);         //Create instance of SonarProxy     2
LaserProxy lp(&robot, 0);         //Create instance of LaserProxy     3
```

Line 1 creates an instance of the PlayerClient object called "robot". Line 2 creates a SonarProxy object called "sp" and associates it with the "robot" object and line 3 does the same for the LaserProxy object "lp". The second constructor parameter for these proxies is an integer representing the index number for the proxy. Here, the indexes are both 0 since there is only one of each.

5.5 Reading and Controlling the Devices

Once devices have been instantiated and added to the robot they can be controlled. The rest of this chapter shows how this is achieved using Player's C++ client library. It shows how to read and control the Pioneer's motors, sonars, laser, bumpers, 5D arm, 2D gripper, camera, pan-tilt-zoom device, virtual blob finder device and ARIA's ACTS blob finder. Note that the standard units for Player are metres, radians and seconds.

5.5.1 The Motors

In order to control the motors it is first necessary to create an instance of a Position2dProxy object and associate it with a PlayerClient object. There are several methods of the Positiond2dProxy class that can be used to set linear and angular speeds and to obtain odometry readings. These are summarised in Table 5.2 below and the following program extract shows how to use some of them. Note that the DTOR() function is specific to the client library and simply converts degrees to radians.

```
double xpos;
double ypos
double zpos;

PlayerClient robot(localhost);     //Create PlayerClient
Position2dProxy pp(&robot, 0);     //Create PositionProxy
pp.SetMotorEnable (true);          //Enable motors

xpos = pp.GetXPos();               //Get position data
ypos = pp.GetYPos();
zpos = pp.GetYaw();

cout << "X pos is " << xpos << "\n";
cout << "Y pos is " << ypos << "\n";
cout << "Z pos is " << zpos << "\n";
for (;;)
                                   //Continuous loop
  {
    pp.SetSpeed(2.0, DTOR(30));    //Set linear speed and turn
  }
```

Table 5.2 Methods of the Position2dProxy class and their functions

Method	Function	Arguments
void SetSpeed()	Sets linear and angular speed (non-holonomic robots)	double aXSpeed = linear velocity double aYawSpeed = angular velocity
void SetMotorEnable()	Enables the motors	bool enable; true = enabled, false = disabled
void ResetOdometry()	Resets odometry to (0, 0, 0)	N/A
void SetOdometry()	Sets odometry to desired position	double aX = desired x co-ordinate double aY = desired y co-ordinate double aYaw = desired angle
double GetXPos()	Gets the x position	N/A
double GetYPos()	Gets the y position	N/A
double GetYaw()	Gets the angular position	N/A

5.5.2 *The Sonar Sensors*

In order to read from the sonar it is first necessary to create an instance of a
SonarProxy object. The program extract below shows how to do this and also
demonstrates use of the GetScan() and GetCount() methods to read the sonar and
determine the number of sonar present. The program continually executes a FOR
loop, which reads the proxy data (using the Read() method of the PlayerClient object), puts the current sonar values into a simple array and drives the robot in a
circle.

```
PlayerClient robot(localhost);              //Create PlayerClient
cout << "You have successfully connected\n";
Position2dProxy pp(&robot, 0);              //Create Position2dProxy
SonarProxy sp(&robot, 0);                   //Create SonarProxy
int numSonar;                               //No. of sonar readings
numSonar = sp.GetCount();                   //Gets no. of sonar readings
double scan_data[numSonar];                 //Array to hold sonar data

for (;;)                                    //Begin read-think-act loop
{
  robot.Read();                             //Reads data for all devices
  for (int i = 0; i < numSonar; i++)        //Loop through sonar readings
  {
    scan_data[i] = sp.GetScan(i);           //Store readings in array
    cout << "sonar data " << i << " " << scan_data[i] << "\n";
  }                                         //End loop through sonar
  pp.SetSpeed(0.1, DTOR(30));               //Move robot slowly in circle
}                                           //End read-think-act loop
```

The method GetPose(x) returns the sonar pose (in m, m, radians) for the sonar
with index number x. Note that individual sonar readings can be referenced using
the syntax sp[0] instead of sp.GetScan(0). Stage 2.0 can simulate the sonar, see
Chapter 6.

5.5.3 *The Laser Sensor*

In order to read from the laser it is first necessary to create an instance of a Laser-
Proxy object. The program extract below shows how to do this and also demonstrates use of the GetRange() and GetCount() methods to read the laser data and get
the number of points on the scan respectively. The program continually executes a
FOR loop, which reads the robot's data, puts the current laser values into a simple
array and drives the robot in a circle.

```
PlayerClient robot(localhost);              //Create PlayerClient
cout << "You have successfully connected\n";
Position2dProxy pp(&robot, 0);              //Create Position2dProxy
LaserProxy lp(&robot, 0);                   //Create LaserProxy
```

```
int numLaserReadings;                      //No. of laser readings
numLaserReadings = lp.GetCount();          //Gets no. of laser readings
double scan_data[numLaserReadings];        //Array to hold laser data

for (;;)                                   //Begin read-think-act loop
{
  robot.Read();                            //Reads data for all devices
  for (int i = 0; i < numLaserReadings; i++) //Loop through laser readings
    {
      scan_data[i] = lp.GetRange(i);       //Store readings in array
      cout << "laser data " << i << " " << scan_data[i] << "\n";
    }                                      //End loop through laser
  pp.SetSpeed(0.1, DTOR(30));              //Move robot slowly in circle
}                                          //End read-think-act loop
```

Table 5.3 below summarises the functions of some of the LaserProxy class methods. The Configure() method allows the user to specify certain laser parameters at run time, such as the maximum and minimum angles that the scan should include and the resolution and intensity values. Note that the resolution and intensity values can also be set by the configuration file, but are overridden when the Configure() method is used. The current settings can be determined by using the GetScanRes(), GetRangeRes(), GetMaxAngle(), GetMinAngle() and GetIntensity() methods. Figure 5.6 shows a set of 361 laser readings and their corresponding angular positions. Note that as with the sonar, individual laser readings can also be referenced using the syntax lp[0] instead of lp.GetRange(0). Stage 2.0 can simulate the laser device, see Chapter 6.

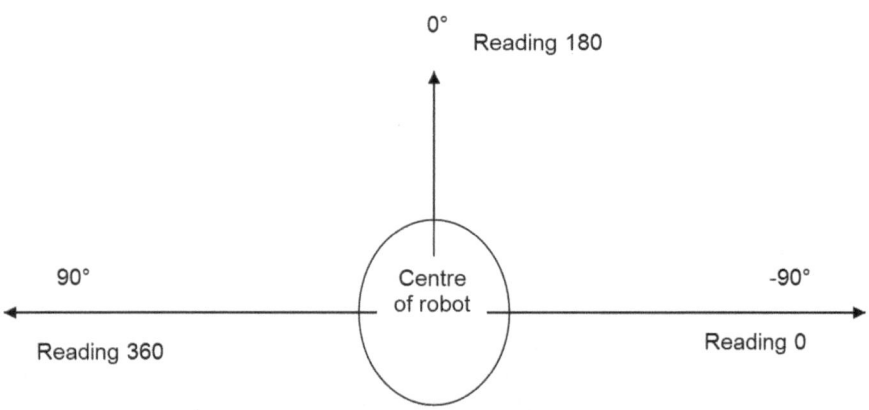

Fig. 5.6 Angular positions of the 361 laser readings

Once the laser and (or) sonar sensors have been read it is useful to write routines that process them in some form or another, for example determining the minimum reading and its angle so that the robot can perform obstacle avoidance if necessary.

Table 5.3 Methods of the LaserProxy class and their functions

Method	Function	Arguments
double GetScanRes()	Get the angular resolution	N/A
double GetRangeRes()	Get the range resolution	N/A
double GetMaxAngle()	Maximum angular position for the scan	N/A
double GetMinAngle()	Minimum angular position for the scan	N/A
double GetRange()	Get a particular laser reading	aIndex = the index of the reading required
double GetCount()	Get the number of points in a scan	N/A
double GetIntensity()	The intensity of a particular laser reading	aIndex = the index of the reading required
void Configure()	For setting laser parameters	aMinAngle = Minimum angle of scan aMaxAngle = Maximum angle of scan aScanRes = Angular resolution aRangeRes = Linear resolution aIntensity = Intensity of scan

The Player C++ client library does not provide any pre-written programs to do this, but an example program is provided as part of the online materials that supplement this book, see the Appendix for more details.

5.5.4 The Bumpers

The bumpers are read by creating an instance of a BumperProxy object and associating it with a PlayerClient object. The program extract below shows how to do this:

```
PlayerClient robot(localhost);         //Create PlayerClient
Position2dProxy pp(&robot, 0)          //Create Position2dProxy
BumperProxy bp(&robot, 0);             //Create BumperProxy
```

In addition, the following program extract shows how a robot named "taylor" may be programmed to respond to the bumpers. The IsAnyBumped() method detects whether any of the bumpers have been activated. The "escapeTraps()" method of a user-written class is called with boolean arguments "cornered" (set to true here if the average laser reading is lower than a threshold value "avTol"), and bp.IsAnyBumped().

```
if (taylor.average <= avTol || bp.IsAnyBumped() == true)
  {
    if (taylor.average <= avTol)
      {
```

```
      cornered = true;
    }else
    {
      cornered = false;
    }
    taylor.escapeTraps(cornered, bp.IsAnyBumped());
}
```

The user-written "escapeTraps()" method is presented below. In this method the angular and linear speeds are dependent both upon whether the robot was cornered and whether it was bumped at the back.

```
void Robot::escapeTraps(bool cornered, bool bumped)
{

int random_number;              //Decide turn direction
int factor;                     //Negative or positive angle
int angle;                      //Turn angle

random_number = (rand()%2);     //Get 0 or 1

if (random_number == 0)
  {
    factor = -1;
  }else
  {
    factor = 1;
  }

if (cornered == true)
  {
    angle = factor * 135;
  }else
  {
    angle = factor * 75;
  }

if (bumped == false)
  {
    steerRobot(-0.1, angle);     //Reverse turn
  }else
  {
    angle = factor * 5;
    cout << "BUMPED AT REAR - MOVING FORWARD BY " << angle << " DEGREES\n";
    steerRobot(0.1, angle);      //Forward turn
  }
}
```

It is also possible to check individual bumpers using the IsBumped() method and supplying the bumper index number as the argument. Note that bp.IsBumped(0) and bp[0] are equivalent where "bp" is a BumperProxy object. Table 5.4 below

summarises some of the functions of the methods of the BumperProxy class. The
bumpers can be simulated in Stage 2.0.4 see Chapter 6.

Table 5.4 Methods of the BumperProxy class and their functions

Method	Function	Arguments
bool IsBumped()	Returns true if specified bumper has been bumped, false otherwise	aIndex = index of bumper
bool IsAnyBumped()	Returns true if any bumper has been bumped, false otherwise	N/A
player_bumper_define_t GetPose()	Returns a specific bumper pose	aIndex = index of bumper
void RequestBumperConfig()	Requests the geometries of the bumper	N/A

5.5.5 The 5D Arm

The drivers for the 5D arm are incorporated into the p2os driver in Player version
2. This adds two new interfaces, an actuator array interface (actarray) that provides
direct control of each joint and an inverse kinematics interface (limb) that allows
specification of a position and orientation for the end effector (the gripper). For a
full explanation of both forward and inverse kinematics see [10]. The configuration
file presented below shows the changes that must be made to include these two new
interfaces. Lines 1 and 2 simply add them to the list of interfaces that the p2os
driver provides, whereas lines 3, 4 and 5 are additional attributes associated with
the limb interface. The limb_pos value specifies the position of the base arm from
the robot centre; this is in radians and the default values are 0, 0, 0. These values
are dependent on the configuration. The limb_links values are the offsets from each
joint to the next in metres. The default settings are 0.06875, 0.16, 0, 0.13775 and
0.11321. The measured distances between each joint are actually 0.06875, 0.16000,
0.092000, 0.04575 and 0.11321. However, the way to specify the link lengths is not
obvious since some of them rotate around different axes to the others. A complete
description of the Pioneer arm and details of how to measure the link lengths are
given in [11]. The limb_offsets values are the angular offsets of each joint from the
desired position to the actual, and the defaults are 0, 0, 0, 0, 0. These values are for
calibration purposes and can be measured by commanding the joints to zero radians
and measuring the actual angle.

```
# Pioneer with arm

driver
(
    name "p2os"
    provides ["odometry:::position2d:0"
    "sonar:0"
    "aio:0"
    "dio:0"
    "power:0"
    "actarray:0"                                                  1
    "limb:0"                                                      2
    "bumper:0"]
    port "/dev/ttyS0"
    limb_pos [0 0 0]                                              3
    limb_links [0.06875 0.16 0 0.13775 0.11321]                  4
    limb_offsets [0 0 0 0 0]                                      5
)
```

The individual joints can be controlled through Player by creating an instance of an ActArrayProxy object and associating it with a PlayerClient object. The ActArrayProxy methods, MoveTo() MoveHome() and MoveAtSpeed() can then be used to command its position (in radians) or speed and the GetCount() method can be used to determine the number of joints. The gripper fingers are controlled by commanding the sixth joint and the GetActuatorGeom() method can be called to determine the angles necessary to close and open them fully. The RequestGeometry() method of the ActArrayProxy class lists each joint and prints out data about its type, centre, minimum, maximum and home positions.

The pose of the end effector can also be controlled by creating an instance of a LimbProxy object, associating it with a PlayerClient object and using the SetPose() method of the LimbProxy class. SetPose() requires full pose information (position, approach vector and orientation vector), where approach is forward from the gripper and orientation is up from the gripper. All values are in metres with x forward, y to the left and z up. If the inverse kinematic calculator cannot find a solution to set the effector to the desired position then the arm's status is PLAYER_LIMB_STATE_OOR (i.e. out of reach). Using lp.SetPose(0.6.671f, 0.0f, 0.309f, 1, 0 0, 0, 0, 1) should stretch out the arm straight in front of the robot. The RequestGeometry() method of the LimbProxy class shows the current limb offset, the end effector position, the approach vector and the orientation vector.

The program below shows how to obtain the number of joints and then set the position of each joint to 20 degrees by using the MoveTo() method of the ActArrayProxy class. Next, the grippers are fully closed and opened using GetActuatorGeom() and its min and max attributes and MoveTo(). The LimbProxy class is then used to extend the arm in front of the robot using the SetPose() method of the class. Finally, the arm is set back to its home position using the MoveHome() method of the ActArrayProxy class. Note that ten calls to the Read() method are necessary in order to obtain the geometry data.

```
#include <libplayerc++/playerc++.h>
#include <unistd.h>
#include <stdlib.h>
#include <iostream>

using namespace std;

int main(int argc, char **argv)
{

using namespace PlayerCc;

/* Connect to Player server */
PlayerClient robot(localhost);

/* Set up arm proxies */
ActArrayProxy ap(&robot, 0);
LimbProxy lp(&robot, 0);

uint numJoints;
int joint = 0;
for (int i = 0; i < 10; i++)
  {
    robot.Read();
  }
numJoints = ap.GetCount();
cout << "There are " << numJoints << " joints\n";

ap.RequestGeometry();
lp.RequestGeometry();

cout << "Position before movement\n";
cout << ap;
cout << lp;

for (joint = 0; joint < numJoints; joint++)
  {
    ap.MoveTo(joint, DTOR(-20));
    sleep(2);
  }

for (int i = 0; i < 10; i++)
  {
    robot.Read();
  }
```

```
lp.RequestGeometry();
cout << "Position after first movement (using MoveTo)\n";
cout << lp;

cout << "Now closing gripper\n";
ap.MoveTo(numJoints-1, ap.GetActuatorGeom (numJoints-1).min);
sleep(2);
cout << "Now opening gripper\n";
ap.MoveTo(numJoints-1, ap.GetActuatorGeom (numJoints-1).max);
sleep(2);

cout << "Setting pose of end effector\n";
lp.SetPose(0.60671f, 0.0f, 0.319f, 1, 0, 0, 0, 0, 1);
sleep(2);
for (int i = 0; i < 10; i++)
  {
    robot.Read();
  }

lp.RequestGeometry();
cout << "Position after second movement (SetPose)\n";
cout << lp;

for (joint = 0; joint < numJoints; joint++)
  {
    ap.MoveHome(joint);
  }

for (int i = 0; i < 10; i++)
  {
    robot.Read();
  }

lp.RequestGeometry();
cout << "Position after homing\n";
cout << lp;

return(0);
}
```

The program produces the output shown below. Note how the position of the end effector changes after each movement.

```
There are 6 joints

Position before movement

6 actuators:
Act Type    Min     Centre    Max     Home    CfgSpd Pos     Speed State Brakes
0   Linear  1.435   -0.000   -1.510  -0.000   0.755  -0.000 0.000 2      N
1   Linear  2.127   -0.000   -1.064  1.963    0.818  0.851  0.000 2      N
2   Linear  1.636   -0.000   -1.554  1.636    0.818  0.851  0.000 2      N
3   Linear  -1.601  0.000     1.525  -0.031   0.763  -0.031 0.000 2      N
4   Linear  -1.441  0.000     1.441  -1.441   0.721  -1.412 0.000 4      N
5   Linear  -1.257  0.000     0.471  0.000    0.785  0.000  0.000 4      N

Limb offset: 0.000, 0.000, 0.000
End effector position: 0.043, -0.003, 0.260
Approach vector: -0.999, -0.030, 0.028
Orientation vector: 0.000, 0.005, -1.000

Position after first movement (using MoveTo)
Limb offset: 0.000, 0.000, 0.000
End effector position: 0.001, -0.003, 0.235
Approach vector: -0.973, -0.030, -0.228
Orientation vector: 0.000, 0.004, -0.974

Setting pose of end effector
Position after second movement (SetPose)
Limb offset: 0.000, 0.000, 0.000
End effector position: -0.046, -0.001, 0.191
Approach vector: -0.846, 0.008, -0.529
Orientation vector: 0.000, -0.020, -0.849

Position after homing
Limb offset: 0.000, 0.000, 0.000
End effector position: 0.297, -0.110, 0.119
Approach vector: -0.264, 0.071, 0.956
Orientation vector: 0.000, 0.329, -0.291
```

The LimbProxy class also has a MoveHome() method that moves the end effector to its home position, and a Stop() method that halts the limb. The arm cannot be simulated in Stage as three dimensions are needed to simulate this device.

5.5.6 The 2D Gripper

The 2D gripper can be read and controlled by creating an instance of a Gripper-Proxy object and associating it with a PlayerClient object. The program extract below shows how to do this:

```
PlayerClient rb(host, port);          //Create PlayerClient
GripperProxy gp(&rb, 0);              //Create GripperProxy
```

The methods of the GripperProxy class are summarised in Table 5.5. In addition, the program extract below demonstrates picking up a puck in the gripper by first checking the state of the break beams. It also goes on to check the state of the beams after the grippers have closed around the puck so that the success of the grab can be determined. Note that the GripperProxy object provides data on the state of the paddles (open or closed) and the beams (clear or broken).

```
uint outerState;                          //State of outer break beam
uint innerState;                          //State of inner break beam
bool havePuck = false;                    //Whether puck held in gripper

for (;;)                                  //Start of read-think-act loop
  {
    robot.Read();
    outerState = gp.GetOuterBreakBeam();  //Read state of outer beam
    innerState = gp.GetInnerBreakBeam();  //Read state of inner beam
    cout << State of gripper  << gp << \n; //Print gripper state

    if (outerState == 1 || innerState == 1) //If either beam broken
      {
        pp.SetSpeed(0.001, 0.0);          //Slow down to grasp puck
        gp.SetGrip(GRIPclose,0);          //Close paddles around puck
        sleep(2);
        gp.SetGrip(LIFTup, 0);            //Lift puck
        sleep(3);
        havePuck = true;                  //Puck successfully collected
      }

    if (havePuck == true && (outerState == 0 && innerState == 0)
      {
        cout << "FAILED TO GRASP PUCK - RE-OPENING PADDLES\n";
        havePuck = false;
        gp.SetGrip(LIFTdown, 0);          //Lower gripper
        sleep(2);
        gp.SetGrip(GRIPopen, 0);          //Release puck
        sleep(2);
      }
  }                                       //End of read-think-act loop
```

The following program extract demonstrates dropping a puck after a successful collection.

```
if (havePuck == true)                     //Collected puck
  {
    gp.SetGrip(LIFTdown, 0);              //Lower gripper
    sleep(3);
    gp.SetGrip(GRIPopen, 0);              //Release puck
    sleep(2);
    pp.SetSpeed(-0.1, 0);                 //Reverse away from puck
    sleep(10);
    havePuck = false;                     //Reset boolean after drop
  }
```

Table 5.5 Methods of the GripperProxy class and their functions

Method	Function
uint GetOuterBreakbeam()	Returns 0 if outer beam not broken, 1 if outer beam broken
uint GetInnerBreakBeam()	Returns 0 if inner beam not broken, 1 if inner beam broken
uint GetPaddlesOpen()	Returns 0 if paddles closed, 1 if paddles open
uint GetPaddlesClosed()	Returns 0 if paddles open, 1 if paddles closed
uint GetPaddlesMoving()	Returns 0 if paddles not moving, 1 if paddles moving
uint GetLiftUp()	Returns 0 if gripper is not raised, 1 if gripper is raised
uint GetLiftDown()	Returns 0 if gripper is not down, 1 if gripper is down
uint GetLiftMoving()	Returns 0 if gripper is not moving up or down, 1 otherwise
void SetGrip()	Sets status of the gripper

The 2D gripper can be simulated in Stage 2.0.4, see Chapter 6.

5.5.7 The Pan-tilt-zoom Camera

By default the Pioneer robots have the pan-tilt-zoom device connected to the AUX port on the P2OS board in the robot. Player cannot control the pan-tilt-zoom unit through this connection so it is therefore necessary to make a cable to connect it to a normal serial port before the canonvcc4 driver can operate. Instructions for making and fitting the cable are given below [3]. Before starting you will need to purchase a VISCA - DB9 conversion cable, a 20-ribbon connection cable and a 20-pin header connector.

- Split the ribbon into two 10-pin sections about half way down the cable. Remove about an inch off pins 10 and 20 to make two 9-pin cable ends. Attach two male DB-9 serial connectors to these ends. The serial connection with pin 1 will go to the serial port on the micro-controller and the other will connect to to the VISCA - DB9 conversion cable. Attach the 20-pin header to the end of the cable that is not separated.
- Remove the top plate and nose from the robot and take out the cable that connects serial ports ttyS0 and ttyS1 of the computer to the serial port of the P2OS. This will be a 20-pin header but will have only 9 wires as the default Pioneer configuration does not use port ttyS1.
- Put the 20-pin header of the new cable into the free serial ports on the computer and connect the wire from ttyS0 (pins 1 to 9) to the serial port on the micro-controller. Connect the other serial connection (pins 11 to 19) to the female DB-9 connector on the VISCA to DB-9 conversion cable. Pass the cable outside the robot cover and replace the nose and top cover of the robot. The other end of the VISCA cable connects to the camera's RS-232C-IN socket (the far right socket).
- To test the connection run /usr/local/Aria/examples/demo on the robot and select 'C' for camera control and '@' for a Canon camera. Select '2'

for the serial port /dev/ttyS1 and then test to make sure that the key controls will pan, tilt and zoom the camera.

Note that the camerav41 driver (responsible for getting the actual camera images) can still operate with the pan-tilt-zoom device connected to the AUX port, so if you only require images and do not need the camera to move then it is not necessary to make a special cable.

A pan-tilt-zoom unit such as the Pioneers' Canon VC-C4 can be controlled by creating an instance of a PtzProxy and associating it with a PlayerClient object. The program extract below shows how to do this:

```
PlayerClient rb(localhost);          //Create PlayerClient
PtzProxy zp(&rb, 0);                 //Create PtzProxy
```

The pan, tilt and zoom values can then be set using the SetCam() method of the PtzProxy class. This method takes three doubles representing pan, tilt and zoom respectively as its arguments. In addition, the current settings for pan, tilt and zoom can be obtained by calling the GetPan(), GetTilt() and GetZoom() methods of the class. The Canon VC-C4 has a pan range of -98 to 98 degrees, a tilt range of -30 to 88 degrees and a zoom range between 0 and 2000 units [1].

For some camera models the SetSpeed() method can be used to set the speeds of the pan, tilt and zoom functions. It takes three double values representing each of these speeds respectively. However, this method does not work for the Canon VC-C4 as it is a very slow device. Some examples of the other methods are shown below:

```
/* Read the current ptz settings */
cout << "Pan " << zp.GetPan() << " Tilt " << zp.GetTilt() << " Zoom "
     << zp.GetZoom() << "\n";

zp.SetCam(DTOR(0), DTOR(88), zp.GetZoom());    //Set pan to 0, tilt to 88
                                               //Leave zoom as it is
```

A generic pan-tilt-zoom camera can be simulated in Stage 2.0., although the tilt function cannot be emulated due to the 2D nature of the simulation, see Chapter 6.

5.5.8 The Virtual Blob Finder Device

This subsection deals with the BlobfinderProxy, which can be used with the generic cmvision driver when using real robots. CMVision (Color Machine Vision) is a blob finding software library written by Jim Bruce, see [9]. Player's cmvision driver includes all of the CMVision source code so there is no need to download and install it separately. The driver provides a stream of camera images to CMVision and assembles the resulting blob information into Player's data format. The driver also requires a camera driver such as camerav41 to obtain the image data. The declaration of these drivers in the configuration file is shown below:

```
driver
(
    name "camerav41"
    provides ["camera:0"]
    port "/dev/video0"
    source 0
    size [320 240]
    norm "ntsc"
    mode "RGB888"
)

driver
(
    name "cmvision"
    requires ["camera:0"]
    provides ["blobfinder:0"]
    colorfile ["colors2.txt"]
)
```

The path to an additional CMVision configuration file should be specified as the argument to the colorfile attribute; here the file is "colors2.txt" in the home directory. The CMVision configuration file has a format as shown below:

```
[Colors]
(255, 128, 0)    0.0000 0 Ball
(255, 255, 0)    0.0000 0 Yellow_Team
(0, 0, 255)      0.0000 0 Blue_Team
(255, 255, 255)  0.0000 0 White
(255, 0, 255)    0.0000 0 Marker_1_(Pink)
(160, 0, 160)    0.0000 0 Marker_2_(Purple)

[Thresholds]
( 12:153,  99:127,143:168)
(101:196,  60:115,114:148)
( 64:157,144:164, 85:129)
(105:192,  68:187,120:131)
( 85:190,  82:189,141:269)
( 30: 85,130:145,135:145)
```

In the colors section the RGB values of the colours to be tracked must be specified. They can also be given a name, for example "Marker_1_(Pink)". In the thresholds section the values are the minimum and maximum tolerances for these colours and are specified in YUV values. If you have a still photograph of an object you need to track you can determine the RGB values using a graphical package such as GIMP. Take readings from a number of pixels and then determine the maximum, minimum and average for each of the R, G and B values. You can then convert the maximum and minimum values to YUV using appropriate formulae. The broadcasting standard CCIR 601 defines the relationship between YUV and RGB as:

$$Y = 0.299R + 0.587G + 0.114B + 0, \qquad (5.1)$$

$$U = -0.169R - 0.331G + 0.500B + 128, \qquad (5.2)$$

$$V = 0.500R - 0.419G - 0.081B + 128, \qquad (5.3)$$

see [8]. You should use the average values (in RGB) for the colors section of the CMVision configuration file and the maximum and minimum values (in YUV) for the thresholds section. The configuration file for the cans described in Section 3.1 is shown below:

```
[Colors]
(158, 81, 74) 1.0000 0 Red


[Thresholds]
(66:189,101:114,160:175)
```

On sampling the pixels the image of the can had an average R value of 158, an average G value of 81 and an average B value of 74. The maximum and minimum R, G and B values were [255, 165, 142] and [112, 48 , 42] respectively. So the lowest Y value is the result of feeding the lowest R, G and B values into Equation 5.1 above, i.e., 66. The highest Y value is the result of feeding the highest R, G and B values into Equation 5.1, i.e. 189. In some cases the number obtained from feeding in the lowest R, G and B values may be higher than the number obtained from feeding in the highest values. In these cases the numbers should still be entered into the thresholds section in numerical order, lowest first.

So far we have described setting up Player and CMVision configuration files in order to use the BlobfinderProxy with a real robot. If you are running your program in Stage you do not need to declare the cmvision and camerav41 drivers in your configuration file or specify a CMVision configuration file. This is because Stage configuration files are written slightly differently to those for real robots. Further details are provided in Chapter 6, which describes setting up a configuration file, world file and robot description file for simulating blob finding in Stage.

Once you have set up your configuration file you can control the virtual blob finding device by declaring an instance of a BlobfinderProxy object and associating it with a PlayerClient object. The program extract below shows how to do this:

```
PlayerClient rb(localhost);        //Create instance of PlayerClient
BlobfinderProxy bp(&rb, 0);        //Create instance of BlobfinderProxy
```

The BlobfinderProxy uses the GetCount() method to return the number of blobs seen and the GetBlob() method (with an unsigned integer as an argument) to specify

a particular blob. Note that bp.GetBlob(0) and bp[0] are equivalent. In addition there
are various attributes to the GetBlob() method. The x attribute shows the blob's
distance along the x axis of the blobfinder window, the area attribute gives the blob's
area and the top attribute indicates how close the blob is to the top of the window.
The color attribute shows the colour of the blob as a packed RGB value and the
GetWidth() and GetHeight() methods of the class return the width and height of the
image respectively.

The program below shows a user-defined method that could be used for determining whether there are blobs of the right colour present and which one should be
tracked, i.e., which has the largest area. Assume that "foundBlob" is a new global
boolean variable for the class that this method is part of, i.e., an attribute of the class.

```
/*
*-------------------------------------------------------------------------
* Use blobfinder to locate pucks of a specified packed RGB value
*-------------------------------------------------------------------------
*/

int Robot::searchPucks(double sd, int rgb, BlobfinderProxy *bfp)
{

int trackIndex;                         //Index of tracked blob
int maxArea = 0;
foundBlob = false;                      //Colour not tested yet
if (bfp->GetCount()>0)                  //If blobs seen

  {
    cout << "THERE ARE " << bfp->GetCount() << " BLOBS.\n";
    for (int j=0;j<bfp->GetCount();j++)         //Loop through blobs
      {
        cout << "blob " << j << "[" << bfp->GetBlob(j).color << "]\n";
        if (bfp->GetBlob(j).color == rgb)       //If colour matches
          {
            foundBlob = true;
            cout << "BLOB " << j << " IS A RED BLOB\n";
            cout << "blob " << j << " x ["<<bfp->GetBlob(j).x<<"]\n";
            cout << "blob " << j << " area ["<<bfp->GetBlob(j).area<<"]\n";
            cout << "blob " << j << " top ["<<bfp->GetBlob(j).top<<"]\n";
            if (bfp->GetBlob(j).area > maxArea) //Find biggest red blob
              {
                maxArea = bfp->GetBlob(j).area;
                trackIndex = j;
              }
          }
      }
  }

}else                                   //If no blobs detected
{
  cout << "NO BLOBS DETECTED!\n";       //No blobs of right colour
  trackIndex=999;
  foundBlob = false;
}
```

```
if (bfp->GetCount() > 0 && foundBlob == false)
  {
    cout << "NO RED BLOBS TO TRACK!\n";          //No blobs detected
    trackIndex=999;
    foundBlob = false;
  }
return (trackIndex);                             //Pass blob index
}
```

The following lines could be used in a main program to call the above method and guide the robot towards the selected blob, if present. Assume that a "Robot" object of a user-defined "Robot" class called "taylor" was declared earlier and that the packed RGB value for tracking is 16711680 (red). Assume also that "wander-Random()" and "steerRobot()" are user-written methods for a random wander and for steering the robot in a given direction respectively.

```
const int RED = 16711680;                        //Packed rgb value
int blobIndex;                                   //Blob to be tracked
double maxSafeSpeed = 0.17;                       //Maximum drive speed

for ;;
  {
    rb.Read();                                   //Read proxies
    if (count%10 == 0)                           //Count one second
      {
        /*Look for red pucks */
        blobIndex = taylor.searchPucks(maxSafeSpeed, RED, &bp);
        cout << "Blob index is " << blobIndex << "\n";

        if (taylor.foundBlob == true)            //If red blob
          {
            cout << "FOUND BLOB\n";
            double turn = (40.0  bp.GetBlob(blobIndex).x)/3.0;
            taylor.steerRobot(maxSafeSpeed, turn);  //Steer robot to blob
          }
        if (taylor.foundBlob == false)           //If no red blobs
          {
            cout << "LOOKING FOR BLOBS\n";
            taylor.wanderRandom(maxSafeSpeed);    //Wander randomly
          }
      }
  }
```

Calling rb.Read() in the above program fills the BlobfinderProxy object with the latest data on all blobs in view. A pointer to the proxy is passed to the user-created "searchPucks()" method of the "Robot" class so that the data can be processed. The method returns the index of the blob with the largest area. If a blob is found the robot turns 40 degrees minus the x value of the blob, which sets the robot moving towards the blob (as the centre of the blob tracking window along the x-axis was set to 40).

The width of the x-axis can be set using the image attribute of the blob finder model in Stage (see Section 6.2) or by setting the size attribute of the camerav41 driver. If the object is central the robot does not adjust its course but if the object is to the right of centre, for example has an x value of 60, then it turns about seven degrees right. Thus the robot is constantly tracking the largest red blob if one is present, otherwise it wanders randomly looking for blobs. The blob finder can be simulated in Stage 2.0.4, see Chapter 6.

5.5.9 Using the Blob Finder with ACTS

You can also use Player's BlobfinderProxy class with MobileRobots' ACTS blob finding software. The ACTS server must be running on a real robot and you must declare the acts driver in your Player configuration file as shown below:

```
driver
(
    name acts
    provides [blobfinder:0]
    configfile /home/amw/actsconfig
    channel 0
)
```

Unlike the cmvision driver there is no need to declare a camera driver for ACTS. The channel attribute refers to the channel or .lut file listed in your ACTS configuration file (see Section 3.1 for further details on training ACTS channels and setting up the configuration file). Once the Player and ACTS configuration files are set up you can use the BlobfinderProxy methods described in Section 5.5.8 to track blobs.

The next chapter looks at Player's 2D simulator Stage and demonstrates how to set up world files and configuration files to describe your virtual robot, its sensors and actuators and its environment.

Chapter 6
Stage Simulations

6.1 Introduction

Stage is a Player plugin module providing simple and computationally cheap two-dimensional emulation of mobile robots and their various devices [2]. Robot control clients that have been written for real robots should be able to run on Stage without requiring any modification and vice versa. The Stage 2.0.4 simulator can model the Pioneer P3-DX differential steer robot base with front and rear sonar, a laser range finder, pan-tilt-zoom unit, virtual blob finder device, bumpers, 2D gripper and odometry.

6.2 Creating World Files

In Stage the robot and its environment are described by creating a world file. The world file declares the window in which the simulation will be displayed, the models it will use and their various attributes. A list of the attributes common to all models and their data types is given below. Note that some attributes require their values to be enclosed in square brackets (these are shown with square brackets in the list), but others do not require any brackets at all.

- pose [float float float] - the pose of the model in its parent's coordinate system.
- size [float float] - the size of the model.
- origin [float float float] - the position of the object's centre relative to its pose.
- color (string) - the colour of the object using a colour name from the X11 database.
- gui_nose (bool) - if 1 draw a nose on the model showing its heading (positive x-axis).
- gui_grid (bool) - if 1 draw a scaling grid over the model.
- gui_movemask (int) - define how the model can be moved by the mouse in the GUI window.

- gui_boundary (bool) - if 1 draw a bounding box around the model indicating its size.
- obstacle_return (bool) - if 1 this model can collide with other models that have this property set.
- blob_return (bool) - if 1 this model can be detected in the blob finder (depending on its colour).
- ranger_return (bool) - if 1 this model can be detected by ranger sensors.
- laser_return (int) - if 0 this model is not detected by laser sensors. If 1 the model shows up in a laser sensor with normal (0) reflectance. If 2 it shows up with high (1) reflectance.
- gripper_return (bool) - if 1 this model can be gripped by a gripper and can be pushed around by collisions with anything that has a non-zero obstacle_return.
- mass (float) - estimated mass in kilogrammes.

The world file may also refer to other files known as include files. These are given the extension .inc and should be listed using the include keyword. The example world file given in this section includes the files "pioneer.inc", which describes a Pioneer robot and "map.inc", which describes a map model.The "pioneer.inc" file is shown first. Note that comments are placed into world and include files by using the # key.

```
# Pioneer p3dx definition

#The laser range finder
define sick_laser laser
(
    range_min 0.0
    range_max 8.0
    fov 180.0 # Field of view
    samples 361 # No of readings each cycle
    color "blue"
    size [0.14 0.14]
)


# The sonar array
define p3dx-sh_sonar ranger
(
    scount 16
    # define the pose of each transducer [xpos ypos heading]
    spose[0] [ 0.115 0.130 90 ]
    spose[1] [ 0.155 0.115 50 ]
    spose[2] [ 0.190 0.080 30 ]
    spose[3] [ 0.210 0.025 10 ]
    spose[4] [ 0.210 -0.025 -10]
    spose[5] [ 0.190 -0.080 -30 ]
    spose[6] [ 0.155 -0.115 -50 ]
```

```
   spose[7] [ 0.115 -0.130 -90 ]
   spose[8] [ -0.115 -0.130 -90 ]
   spose[9] [ -0.155 -0.115 -130 ]
   spose[10] [ -0.190 -0.080 -150 ]
   spose[11] [ -0.210 -0.025 -170 ]
   spose[12] [ -0.210 0.025 170 ]
   spose[13] [ -0.190 0.080 150 ]
   spose[14] [ -0.155 0.115 130 ]
   spose[15] [ -0.115 0.130 90 ]

   # define the field of view of each transducer

   # [range_min range_max view_angle]

   sview [0 5.0 15] # min and max range in metres

   # define the size of each transducer [xsize ysize] in metres
   ssize [0.01 0.05] # size in metres - has no affect on data
)

define p3dx-sh_gripper gripper
(
   pose [0.22 0 0]
   color "black"
)

define p3dx-sh_blobfinder blobfinder
(
   channel_count 6
   channels ["red" "blue" "green" "cyan" "yellow" "magenta"]
   range_max 8.0
   image [80 60]
)

define p3dx-sh_camera ptz
(
   size [0.08 0.08]
   ptz [0.0 0.0 60.0]
   ptz_speed [1.0 0.0 0.3]
   p3dx-sh_blobfinder()
)

#define 5 straight bumpers around the rear edge of the robot
define p3dx-sh_bumper bumper
(
```

```
   bcount 5

   bpose[0]  [ -0.25 0.22 128 0.105 0.0 ]
   bpose[1]  [ -0.32 0.12 161 0.105 0.0 ]
   bpose[2]  [ -0.34 0.00 180 0.105 0.0 ]
   bpose[3]  [ -0.32 -0.12 199 0.105 0.0 ]
   bpose[4]  [ -0.25 -0.22 232 0.105 0.0 ]
)


# The configuration
define p3dx-sh position
(
   # actual size
   size [0.445 0.400]

   # the pioneer's center of rotation is offset from its center of area
   origin [-0.04 0.0 0]

   # draw a nose on the robot so we can see which way it points
   gui_nose 1

   # estimated mass in KG
   mass 15.0

   # this polygon approximates the shape of a pioneer
   polygons 1
   polygon[0].points 8
   polygon[0].point[0]  [ 0.23 0.05 ]
   polygon[0].point[1]  [ 0.15 0.15 ]
   polygon[0].point[2]  [ -0.15 0.15 ]
   polygon[0].point[3]  [ -0.23 0.05 ]
   polygon[0].point[4]  [ -0.23 -0.05 ]
   polygon[0].point[5]  [ -0.15 -0.15 ]
   polygon[0].point[6]  [ 0.15 -0.15 ]
   polygon[0].point[7]  [ 0.23 -0.05 ]

   # Pioneers have differential steering
   drive "diff"

   obstacle_return 1
   p3dx-sh_sonar()
   p3dx-sh_gripper()
   p3dx-sh_camera()
   sick_laser()
)
```

The above file is used to describe the sensor and actuator configuration of a Pioneer robot. The define statement is used to create a new model that can be referred to later in the file and also in the world file. The first model created is of type laser and is given the name "sick_laser". The additional attributes of laser models are:

- samples (int) - the number of readings in each cycle, i.e. the number of laser samples per scan.
- range_min (float) - the minimum range reported by the scanner in metres. The scanner will detect objects closer than this but report their range as the minimum.
- range_max (float) - the maximum range reported by the scanner in metres. The scanner will not detect objects beyond this range.
- fov (float) - the maximum angular field of view of the scanner in degrees.

In the file shown all of the above are set and two attributes generic to all models are also set; color and size. Next the sonar arrangement is described by creating a model called "p3dx-sh_sonar" of type ranger. The additional attributes of ranger models are:

- scount (int) - the number of range transducers.
- spose [<transducer index>] [float, float, float] - pose of the transducer relative to its parent.
- ssize [float float] - size in metres. This has no effect on the data, it only determines how the sensor looks in the Stage window.
- sview [float float float]- minimum range and maximum range in metres, field of view angle in degrees.

In the sample file the spose values are set for each of the sixteen transducers and the ssize and sview values are also set. The next model created is of type gripper and is given the name "p3dx-sh_gripper". There are no additional attributes for gripper models. The fourth model created is of type blobfinder and is given the name "p3dx-sh_blobfinder". Blobfinder models have the additional attributes:

- channel_count (int) - the number of channels, i.e., the number of discrete colours detected.
- channels[string string ...] - this list defines the colours detected in each channel using colour names from the X11 database. The number of strings must match channel_count.
- image[int int] - the width and height of the image in pixels. This determines the blob finder's resolution.
- ptz[float float float] - sets the pan, tilt and zoom angle (fov) of the blob finder although tilt angle has no effect.
- range_max float - maximum range of the sensor in metres.

In the example, a window size of 80 x 60 pixels and six channels are set and the blobfinder model is set to detect objects as far away as eight metres. A model of type

ptz, called "p3dx-sh_camera" is created next. Ptz models share the ptz attribute of
blobfinder models and also have the additional attribute ptz_speed [float float float],
which controls the speed at which the pan, tilt and zoom angles are set. Note how
the "p3dx-sh_blobfinder" model is added to the "p3dx-sh_camera" model to make
the blob finder device a child of the ptz device. This means that the pan, tilt and
zoom of the blob finder are set by controlling the pan, tilt and zoom of the camera.

The final sensors declared are the bumpers. The attribute bcount is used to set
the number of bumpers, and then the pose of each bumper is expressed using the
bpose<bumper index> attribute.

The last model created is called "p3dx-sh" and is of type position. This defines
the whole robot including all of its sensors and actuators. Here, the generic model
attributes size, origin, gui_nose and mass are used to describe the robot and a model
type called polygons is used to describe the shape of the robot. One polygon with
eight points is declared using the polygons<index number>.points attribute. The
co-ordinates for each point of the polygon are then set using the polygons<index
number>.point attribute. Position models also have a drive attribute which can be
set to diff, omni or car, depending on whether the robot uses a differential-steer
mode (Pioneers), omni-directional mode or is car-like. The devices created earlier,
the sonar, laser, gripper and camera (which includes the blob finder as a child) are
attached to the robot in the final lines of the "p3dx-sh" declaration section.

The "map.inc" file defines a model called "map" that is of generic type model.
This sets some basic properties of the plan of the robot's environment including how
the robot's sensors will respond to it. For example, here the gripper_return attribute
is set to 0 as the robot should not treat walls etc. as objects to be grasped and pushed.
The file is presented below:

```
define map model
(
    color "black"
    boundary 1
    gui_nose 0
    gui_grid 1
    gui_movemask 0
    gui_outline 0
    gripper_return 0
)
```

The example world file that includes the "map.inc" and "pioneer.inc" files is presented below:

```
# Test world for Pioneer p3dx

# Include file that defines Pioneers
include "pioneer.inc"
```

```
# Include 'map' object used for floorplans
include "map.inc"

# size of the world in metres
size [15 15]

# speed of the simulation
# interval_sim 100 # milliseconds per update step
# interval_real 50 # real-time milliseconds per update step
# gui_interval 100 # rate at which window is re-drawn (milliseconds)

# set the resolution of the underlying raytrace model in metres
resolution 0.02

# configure the GUI window
window
(
   size [ 510.000 750.000 ] # size of window
   center [0.006 -0.009]
   scale 0.008 # size of each bitmap pixel in metres
)

# load an environment bitmap - this is defined in map.inc
map
(
   bitmap "pen.png"
   size [3 5] # size of the bitmap drawing
   map_resolution 0.02
   name "pen"
)

# create a robot
p3dx-sh
(
   name "robot1"
   color "red"
   pose [-0.166 -1.177 86.295]
)

define puck model
(
   size [ 0.08 0.08 ]
   gui_movemask 3
   gui_nose 0
```

```
    ranger_return 0
    laser_return 0
    blob_return 1
    gripper_return 1
    obstacle_return 1
)
puck( pose [-1.114 1.467 -105.501 ] color "red" )
puck( pose [-0.781 -0.264 -307.877 ] color "blue" )
puck( pose [-1.095 0.838 -461.643 ] color "green" )
```

The world entity has attributes:

- interval_sim (milliseconds) - the length of each simulation update cycle in milliseconds.
- interval_real (milliseconds) - the amount of real-world (wall-clock) time the simulator will attempt to spend on each simulation cycle.
- resolution (metres) - the resolution of the underlying bitmap model. Larger values speed up raytracing at the expense of fidelity in collision detection and sensing.
- size [float float] - the size of the world in metres.
- gui_interval (milliseconds) - the rate at which the Stage window is redrawn.

The interval_sim and interval_real attributes can be used to specify how fast the Stage simulation should run in relation to real time. When these are the same Stage runs at approximately real time, but if interval_real is set to less than interval_sim it runs faster than real time. If interval_real is set to 0 Stage runs as fast as it can. The attribute gui_interval is set at 100 by default, i.e., the Stage window is re-drawn every 100 milliseconds. If the value is increased from 100 then the window is redrawn less frequently, which enables Stage to run even faster in some circumstances (see Section 6.5 for further details about accelerated simulations).

After setting the world properties a window entity is declared to describe how the Stage window will be drawn on screen. The window entity has attributes center, size and scale. Size sets the window size in pixels, center sets the location of the centre of the window in world coordinates (metres) and scale sets the ratio of the world to the pixel coordinates i.e., the window zoom. A map model is then declared, which was defined earlier in the "map.inc" file. Within this you must specify the path to the actual image that contains the floor plan by using the bitmap attribute. Images must be drawn in black and white and saved either as bitmaps (.bmp), portable network graphics (.png) or portable pix map (.ppm) files. The "pen.png" image, which represents the real robot pen shown in Figure 5.1, is shown in Figure 6.1.

A "p3dx-sh" model, which was defined earlier in the "pioneer.inc" file is declared next. It is given the name "robot1" and assigned a color and pose. Finally, another model type is defined called "puck". This could also be defined in another include file, but here it is defined within the world file itself. The "puck" objects are created for detection using the blob finder and the robot must be able to grasp them in its gripper and transport them. It must be able to push the pucks and they should not be detected by the laser or sonar. Thus, the laser_return and ranger_return values are

Fig. 6.1 A Stage floor plan

set to 0 and the gripper_return and blobfinder_return values are set to 1. It would also be useful to be able to move the pucks in Stage by left-clicking the mouse and dragging them, so the gui_movemask attribute is given value 3. After having defined the "puck" model, three are declared in the final lines of the file and given a colour and pose.

6.3 Creating Configuration Files

A configuration file also needs to be created before Player can run with Stage. It takes a similar format to the configuration file used with real robots, but needs to specify the stage driver. In fact, the driver must be listed at least twice, first as providing the simulation (and here the world file must also be given), and then as providing all the other functions, for example position2d, laser, sonar, ptz, blobfinder and gripper. The second listing should also include the name of the robot that these functions apply to (given in the world file). An example Stage configuration file is shown below.

```
# Configuration file for controlling Stage devices

# load the Stage plugin simulation driver
driver
(
```

```
   name "stage"
   provides ["simulation:0" ]
   plugin "libstageplugin"

   # load the named file into the simulator
   worldfile "simple.world"
)

driver
(
   name "stage"
   provides ["position2d:0" "laser:0" "sonar:0" "ptz:0"
                                         "blobfinder:0" "gripper:0"]
   model "robot1"
)
```

In the world file the "p3dx-sh" robot declared was given the name "robot1". In the configuration file above the name "robot1" is used to indicate that all robots with this name should be given these functions. Note that multiple names can be used. For example, we might have declared a "p3dx-sh" robot with the name "robot2". If we wanted this robot to have only the laser, sonar and positional functionality we would add an additional five lines to the configuration file as shown below:

```
driver
(
   name "stage"
   provides ["position2d:0" "laser:0" "sonar:0"]
   model "robot2"
)
```

6.4 Running Stage

Stage 2.0 is not a binary and therefore cannot be run independently of Player. To run it the Player binary is called along with the configuration file that specifies the stage driver, see Section 6.3. For example, if you have a Stage configuration file called "mysimple.cfg" in your home directory, you are in your home directory and your system knows the path to the Player binary, you would type the following:

```
player mysimple.cfg.
```

If the system does not know the Player path then you will need to type the full path.

Figure 6.2 below shows the Stage GUI. The *File* menu provides access to the sub-menus *Save*, *Reset*, *Screenshot* and *Exit*. The *Save* function writes the current robot positions, object positions and magnifications into a world file. This is saved under the same name and in the same location as the world file currently open. The *Reset* function reloads the world file so that the robot and objects are set back to their starting positions. The *Screenshot* function allows the user to save the current world as an image, either a jpeg or a png file. In addition, a single frame or a sequence of frames (a movie mpeg) may be saved. *Exit* quits the application. The right mouse button is used for zooming in and for turning robots and objects and the left mouse button is used to grab and move robots and objects and pan left and right and up and down.

The *Edit* menu provides access to a sub-menu *Preferences*, a new feature added to Stage version 2.0 that allows the user to change the values for interval_sim, interval_real and gui_interval in the world file at run time, see Section 6.2 . Figure 6.3 shows the dialogue box that is displayed when this feature is used.

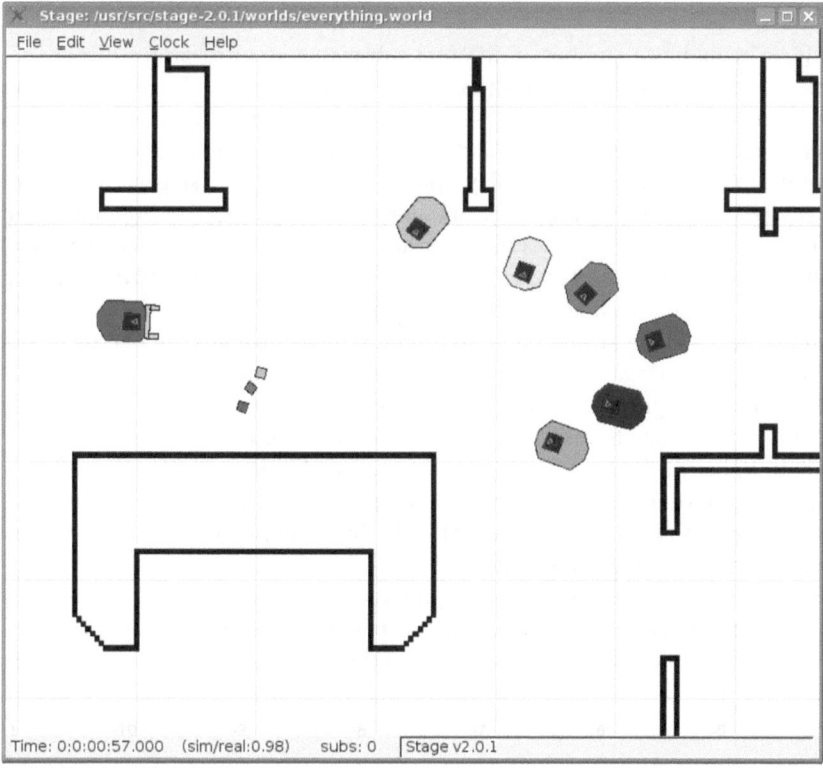

Fig. 6.2 The Stage GUI

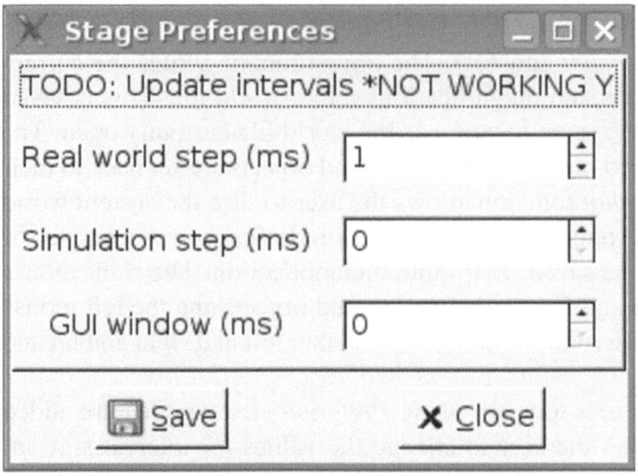

Fig. 6.3 Stage preferences dialogue box

The *View* menu has a number of different sub-menus to turn various display parameters off and on. These include showing the robot's trail (see Figure 6.7, where
the trail is shown in red), displaying the grid lines and printing the position text on
the window, i.e. current velocity, x, y and z co-ordinates, see Figure 6.6. In addition,
there are data and configuration display options for most of the sensors, for example, *laser data* shows the current laser output shaded in purple, see Figure 6.6 and
Figure 6.9, and *laser config* shows the laser field of view, see Figure 6.6. This also
shows the blob finder field of view, which is selected by checking *blob config* on
the menu. The gripper beam display can also be turned off and on by checking *gripper data* on the *View* menu, see Figure 6.8. Figure 6.4 illustrates the pan-tilt-zoom
and laser configurations and Figure 6.5 shows the sonar configuration. (Figures 6.4
to 6.9 are all screen grabs from Player 2.0.1 running Stage 2.0.1, which uses the
configuration, world and bitmap files presented in Section 6.2 and Section 6.3.)

The bottom bar of the GUI displays time data, see Figure 6.2. The elapsed time
since Stage began running is shown in the far left hand corner and the ratio of real
to simulated time is given to the right of this. In this example both interval_real and
interval_sim are set to the same value in the world file, so the ratio is 0.98. This
means that the simulation is running in approximately real time. The *Clock* menu
has only one sub-menu *Pause*; checking this suspends any executing program. When
it is unchecked again the program resumes as if there had been no discontinuation.

6.5 Accelerated Simulations

In the Player examples supplied online as part of the materials to support this book
(see Appendix A), simulations up to 10 times faster than real speed have been

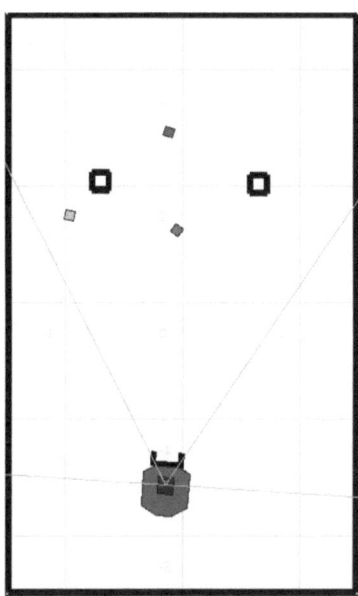

Fig. 6.4 Stage simulation showing ptz and laser configurations

achieved using Stage 2.0.1. The world file had interval_sim set to 100 and interval_real set to 10. Experiments were done setting interval_real to 0 and increasing the value of the gui_interval setting, but these produced no further increase in speed. Delays in the "gripper.cc" and "griptrack.cc" programs were originally set using the unix sleep() function, which takes its argument in seconds. However, when running simulations faster than real speed this caused problems as the commands were still carried out in real time despite the Stage world file settings, i.e., sleep(1) still caused the system to wait 1 second, even when Stage was running 10 times faster. To get around this problem the sleep() commands were changed to usleep() commands, which take their arguments in microseconds and the arguments were varied inversely with the speed of the simulation.

In addition to the problem described above, the timings of the read-think-act loops also caused difficulties when running the "goalseek.cc", "gripper.cc" and "griptrack.cc" programs faster than real time. Commands need to be carried out approximately every second because the robot is given angular speeds for goal seeking and blob tracking, rather than fixed angles to turn through. The line:

```
if (count%10 == 0)
```

in the main read-think-act loop performs this function as messaging runs at approximately 10 Hz. This ensures that commands are carried out only on every tenth execution of the loop, i.e., approximately every second. However, when fast sim-

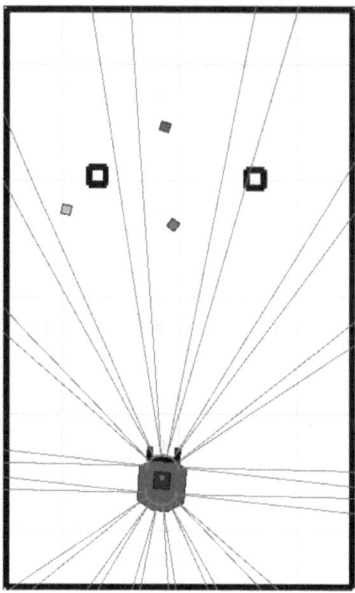

Fig. 6.5 Stage simulation showing sonar configuration

ulations were used it was necessary to reduce the counter value inversely with the speed of the simulation. The formula:

```
int(round(10/simSpeed))
```

was used as the counter value where "simSpeed" was the speed of the simulation. Simulations 1, 2, 4 and 10 times faster than real speed were performed using all three programs with the robot turning towards the goal (and the blobs in the case of "griptrack.cc") adequately in all cases. The formula:

```
int((2*round(10/simSpeed)))
```

was used in the case of "goalseek.cc", except where "simSpeed" was equal to 1.

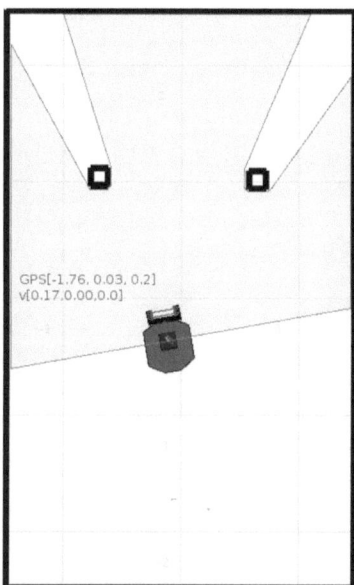

Fig. 6.6 Stage simulation showing laser output and position text

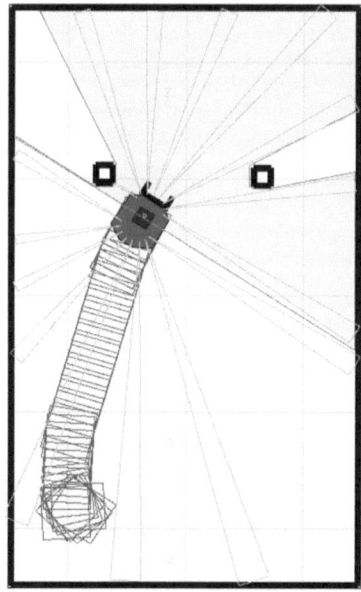

Fig. 6.7 Stage simulation showing laser and sonar outputs and the trail of the robot

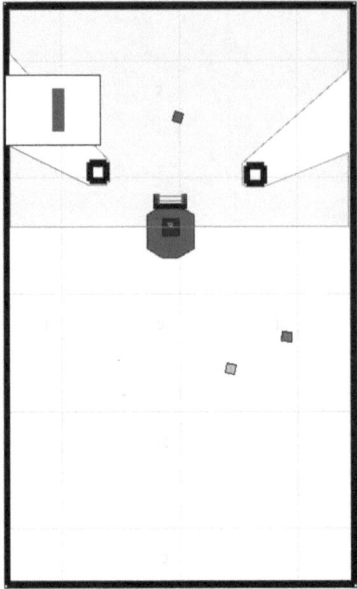

Fig. 6.8 Stage simulation showing laser, gripper beam and blob finder outputs

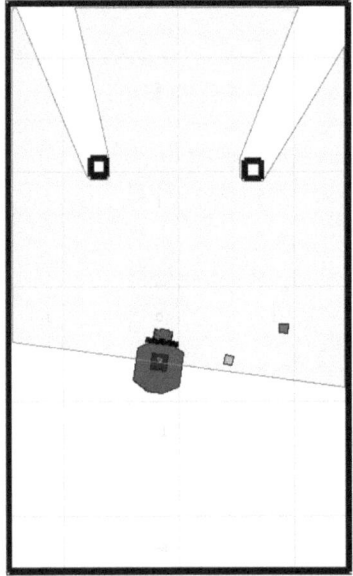

Fig. 6.9 Stage simulation showing laser output. The robot is seen transporting a puck in its grippers

Appendix A
Guide to the Extra Materials

A.1 Folders

The online materials (http://extras.springer.com) that supplement this book are divided into 2 folders, ARIA and Player. The ARIA folder contains a control program "control.cpp" that uses a number of simple behaviours contained in the files "RobotModes.cpp" and "RobotModes.h". These behaviours do not inherit from the ArAction class. The control program also uses data processing techniques found in the files "SensorData.cpp" and "SensorData.h". The control program can perform various demos depending on which macros are defined at the beginning of the program. For example, if you want to run the arm movement demo, uncomment the ARM definition and recompile. The ARIA folder also contains a single blob finding action (the file "BlobFind.cpp" is the action and "single_control.cpp" is the file that uses it), a group blob finding action (the file "BlobFindGroup.cpp" is the action group and "group_control.cpp" is the file that uses them) and a blob finding mode (the file "BlobFindMode.cpp" is the mode and "mode_control.cpp" is the file that uses it). The ARIA folder also icludes a bash script "acomp" for quick compilation of ARIA client programs, and a test ACTS configuration file "actsconfig" with its associated .lut files "channel1.lut" and "channel2.lut". A subfolder Images comprises images of a test robot environment and cans for ACTS training. It also includes a test map for use with MobileSim.

The Player folder contains a configuration file for a real robot "config-Player2.cfg" and world, include and configuration files for Stage 2 simulations "simple.world", "mysimple.cfg", "map.inc", "sick.inc" and "pioneer.inc". There are four demo programs, "goalseek.cc", which performs goal seeking in the world described by "simple.world", "gripper.cc", where the robot collects blocks and transports them through the goal, "griptrack.cc", where the robot tracks only the red blocks and transports them through the goal, and "joint.cc" which controls the 5D arm. The first three programs make use of behaviours and data processing methods contained in the files "Robot.cpp" and "Robot.h". "WorldReader.h" is used with "goalseek.cc" when simulated robots are required to go to specific co-ordinates in

109

the world rather than find the gate and travel through it. It simply reads in the starting position of the robot from the world file. User documentation for these user-written "Robot" and "WorldReader" classes is available in [13]. The "joint.cc" program requires the file "args.h", a standard test file supplied with Player. The files "miniRobot.h", "miniRobot.cpp" and "minigoal.cc" are more simplified versions of "Robot.h", "Robot.cpp" and "goalseek.cc". The Player folder also includes a bash script "pcomp" for quick compilation of Player client programs and a subfolder bitmaps with the image "pen.png" needed for the test Stage simulations.

A.2 Testing the Programs

The programs in the ARIA folder were tested using a Pioneer P3-DX robot running Debian Linux 2.6.10 with ARIA version 2.4.1 and ACTS version 2.2.1. The programs in the Player folder were tested by running the Player server on the same robot with Player 2.0.1 installed in the default location. The client programs were run on a remote PC running Debian Linux 2.6.10 with Player 2.0.1 also installed in the default location. The program "joint.cc" was tested on a different Pioneer P3-DX robot with a 5D arm, running the same operating system and with Player 2.0.1 installed in the default location. The programs "goalseek.cc", "gripper.cc" and "griptrack.cc" were tested in simulation using the remote PC described and Stage 2.0.1 installed in the default location. The program "joint.cc" was not tested in simulation as Stage does not support the 5D arm device.

References

1. Gerkey, B.: The Player Robot Device Interface (2005)
   http://playerstage.sourceforge.net/doc/Player-2.0.0/player/index.html. Cited 3rd June 2009
2. Vaughan, R.: The Stage Robot Simulator (2007)
   http://playerstage.sourceforge.net/doc/Stage-2.0.0/. Cited 3rd June 2009
3. Bryant, J. L.: Instructions for Rewiring a Pioneer Robot so that the PTZ Camera Device can
   be Connected to a Serial Port (ttyS1) on the On-board Computer (2006)
   http://playerstage.sourceforge.net/index.php?src=faq#evid30_wiring. Cited 3rd June 2009
4. Vaughan, R. T., Gerkey, B., Howard, A.: On Device Abstractions For Portable, Reusable
   Robot Code. In: Proceedings of the IEEE/RSJ International Conference on Intelligent Robot
   Systems, pp. 2121-2427, Las Vegas, USA (2003)
5. Gerkey, B., Vaughan, R. T., Howard, A.: The Player/Stage Project: Tools for Multi-Robot
   and Distributed Sensor Systems. In: Proceedings of the 11th International Conference on
   Advanced Robotics, pp. 317-323, Coimbra, Portugal (2003)
6. Gerkey, B. P., Vaughan, R. T., Sty, K., Howard, A., Sukhatme, G., S., Mataric, M. J.:
   Most Valuable Player: A Robot Device Server for Distributed Control. In: Proceedings of
   the IEEE/RSJ International Conference on Intelligent Robots and Systems, pp. 1226-1231,
   Wailea, Hawaii (2001)
7. SourceForge.Inc.: SourceForge.net email archive: PlayerStage-users (2009)
   http://sourceforge.net/mailarchive/forum.php?forum_name=playerstage-users. Cited 3rd June
   2009
8. Wilson, D.: RGB/YUV Pixel Conversion (2007)
   http://www.fourcc.org/fccyvrgb.php#BOURGEOIS. Cited 3rd June 2009
9. Bruce, J.: CMVision Realtime Color Vision (2006)
   http://www.cs.cmu.edu/ jbruce/cmvision/. Cited 3rd June 2009
10. Niku, S.B.: Introduction to Robotics: Analysis, Systems, Applications. Prentice Hall (2001)
11. Gan, J, Q., Oyama, E., Rosales, E. M., Hu, H.: A complete analytical solution to the inverse
    kinematics of the Pioneer 2 robotic arm. Robotica **23**, 123-129 (2005)
12. Oates, R.: Bob's Guide To using ARIA (2006)
    http://www.boboates.co.uk/aria.pdf. Cited 3rd June 2009
13. Whitbrook, A. M.: An Idiotypic Immune Network for Mobile Robot Control. MSc Disserta-
    tion, School of Computer Science, University of Nottingham (2005)
14. ActivMedia Robotics Interface for Application (ARIA) ActivMedia, NH, USA, (2005), avail-
    able with ARIA software download
15. Pioneer 3-SH Operations Manual, Version 1, ActivMedia, NH, USA, (2004)
16. ACTS User Manual, Version 6, ActivMedia, NH, USA, (2006)
17. ActivMedia Robotics' Pioneer Gripper Manual, Version 6, ActivMedia, NH, USA, (2004)
18. ActivMedia Robotics' Pioneer Arm Manual, Version 5, ActivMedia, NH, USA, (2004)

19. Laser Range-Finder Installation and Operations Manual, Version 1, ActivMedia, NH, USA, (2002)
20. PTZ Robotic Cameras Manual, Version 3, ActivMedia, NH, USA, (2003)
21. VC-C4/VC-C4R Instruction Manual, Version 3, Canon, Japan (2000)

Index